Probabilistic risk assessment of engineering systems

Probabilistic risk assessment of engineering systems

Mark G. Stewart
and
Robert E. Melchers

*Department of Civil, Surveying
and Environmental Engineering,
University of Newcastle,
NSW, Australia*

CHAPMAN & HALL
London · Weinheim · New York · Tokyo · Melbourne · Madras

Published by Chapman & Hall, 2–6 Boundary Row,
London SE1 8HN, UK

Chapman & Hall, 2–6 Boundary Row, London SE1 8HN, UK

Chapman & Hall GmbH, Pappelallee 3, 69469 Weinheim, Germany

Chapman & Hall USA, 115 Fifth Avenue, New York, NY 10003, USA

Chapman & Hall Japan, ITP-Japan, Kyowa Building, 3F,
2-2-1 Hirakawacho, Chiyoda-ku, Tokyo 102, Japan

Chapman & Hall Australia, 102 Dodds Street, South Melbourne,
Victoria 3205, Australia

Chapman & Hall India, R. Seshadri, 32 Second Main Road, CIT East,
Madras 600 035, India

First edition 1997

© 1997 Mark G. Stewart and Robert E. Melchers

Typeset in 10/12pt Palatino by Florencetype Ltd, Stoodleigh, Devon

Printed in Great Britain by St Edmundsbury Press Ltd,
Bury St Edmunds, Suffolk

ISBN 0 412 80570 7

Apart from any fair dealing for the purposes of research or private study, or criticism or review, as permitted under the UK Copyright Designs and Patents Act, 1988, this publication may not be reproduced, stored, or transmitted, in any form or by any means, without the prior permission in writing of the publishers, or in the case of reprographic reproduction only in accordance with the terms of licences issued by the Copyright Licensing Agency in the UK, or in accordance with the terms of licences issued by the appropriate Reproduction Rights Organization outside the UK. Enquiries concerning reproduction outside the terms stated here should be sent to the publishers at the London address printed on this page.

The publisher makes no representation, express or implied, with regard to the accuracy of the information contained in this book and cannot accept any legal responsibility or liability for any errors or omissions that may be made.

A catalogue record for this book is available from the British Library

Library of Congress Catalog Card Number: 97-69704

∞ Printed on permanent acid-free text paper, manufactured in accordance with ANSI/NISO Z39.48-1992 and ANSI/NISO Z39.48-1984 (Permanence of Paper)

Contents

Preface x

Chapter 1 – Introduction 1
  1.1 Risk assessment 1
  1.2 Risk assessment in decision-making 2
  1.3 Defining risk 3
      1.3.1 Risk versus probability 3
      1.3.2 Types of risk 4
  1.4 Risk-based decision process 5
      1.4.1 Context definition 5
      1.4.2 Criteria define 7
      1.4.3 Hazard identification 7
      1.4.4 Risk analysis 8
      1.4.5 Sensitivity analysis 9
      1.4.6 Presentation of results 9
      1.4.7 Risk assessment/criteria 9
      1.4.8 Risk treatment 10
      1.4.9 Monitoring and review 11
  1.5 Some other issues 11
      1.5.1 Managerial commitment 11
      1.5.2 Documentation and quality assurance 12
      1.5.3 Politics and public participation 12
      1.5.4 High consequence-low probability events 13
      1.5.5 Legal interpretations and risk 13
  1.6 Conclusion 14
      References 15

Chapter 2 – Sources of risk 16
  2.1 Introduction 16
  2.2 Sources of risk for specific engineering systems 17
      2.2.1 Nuclear power plants 17
      2.2.2 Chemical and oil process and storage plants 21

Contents

	2.2.3 Shipping	23
	2.2.4 Pipelines	24
	2.2.5 Structures	25
	2.2.6 Offshore structures	27
	2.2.7 Dams	28
	2.2.8 Hydrological systems	29
	2.2.9 Aerospace systems	29
	2.2.10 Industrial robots	31
	2.2.11 Computer systems	32
2.3	Unforeseen events	32
2.4	System complexity	34
2.5	Recovery from failure	35
2.6	Summary	36
	References	37

Chapter 3 – Modelling of systems — 40

3.1	Introduction	40
3.2	System failure	42
3.3	Identification of sources of risk	43
	3.3.1 PHA – Preliminary Hazard Analysis	44
	3.3.2 FMEA – Failure Modes and Effect Analysis	46
	3.3.3 FMECA – Failure Mode, Effect and Criticality Analysis	49
	3.3.4 HAZOP – Hazard and Operability studies	50
	3.3.5 Incident databanks	51
3.4	System representation	54
	3.4.1 General approach	54
	3.4.2 Fault trees	56
	3.4.3 Event trees	60
	3.4.4 System dependency – common cause failures	65
3.5	System element performance	69
	3.5.1 Quantitative description	69
	3.5.2 Variables as point estimates	70
	3.5.3 Random variables	70
3.6	Uncertainty in analysis outcomes	74
3.7	Summary	75
	References	75

Chapter 4 – Performance of system elements — 78

4.1	Introduction	78
4.2	Reliability and failure data	79
	4.2.1 Description of data	79
	4.2.2 Types of reliability and failure data	80
	4.2.3 Overall failure rates	80

	4.2.4	Failure rates in individual failure modes	81
	4.2.5	Variation of failure rates	81
	4.2.6	Unavailability	82
	4.2.7	Repair times	83
4.3	Sources of failure data		83
	4.3.1	Experimental data	83
	4.3.2	Expert opinion	86
	4.3.3	Bayes Theorem – combining different data	88
	4.3.4	Reliability databases	90
	4.3.5	Influence of external factors	95
4.4	Reliability of load-resistance sub-systems		97
	4.4.1	Modelling of resistances	100
	4.4.2	Models for material and geometric variables	100
	4.4.3	Derived models for resistance	104
	4.4.4	Modelling of loads	105
4.5	Modelling of consequences		115
4.6	Summary		117
	References		117

Chapter 5 – Human error and human reliability data   122

5.1	Introduction		122
5.2	Human error and human behaviour		124
	5.2.1	Definition of human error	124
	5.2.2	Classification of human error	124
	5.2.3	Performance Shaping Factors	129
	5.2.4	Violations	130
	5.2.5	Unforeseen errors	131
	5.2.6	Error control	132
5.3	Human reliability data		134
	5.3.1	Sources of human reliability data	134
	5.3.2	Human reliability databases	136
	5.3.3	Validation of human reliability databases	147
5.4	Summary		148
	References		149

Chapter 6 – System evaluation   154

6.1	Introduction		154
6.2	Frequentist data, probabilities and uncertainties		155
6.3	Qualitative risk analysis		157
6.4	Quantified Risk Analysis (QRA)		158
	6.4.1	Event trees	159
	6.4.2	Fault trees	161
	6.4.3	Integrated systems	162
6.5	Probabilistic Risk Analysis (PRA)		164

	6.5.1	Overview	164
	6.5.2	Second moment analysis	165
	6.5.3	Full distributional analysis: Monte Carlo simulation	170
	6.5.4	Simplification of the system	173
6.6	System dependency		174
	6.6.1	Explicit methods	174
	6.6.2	Implicit methods	175
	6.6.3	Implicit methods: system reliability cut-off method	176
	6.6.4	Implicit methods: common cause failure methods	176
6.7	Effect of time		179
	6.7.1	Effect of time on failure rates	180
	6.7.2	Time to first failure for components	181
	6.7.3	Time to failure and probability of failure for systems	183
6.8	Reliability of load-resistance sub-systems		185
	6.8.1	Basic concepts	185
	6.8.2	Simple formulation	186
	6.8.3	Generalised formulation	189
	6.8.4	First Order Second Moment (FOSM) method	190
	6.8.5	Sub-systems and bounds	194
	6.8.6	First Order Reliability method	196
	6.8.7	Monte Carlo simulation methods	197
6.9	Sensitivity analysis and updating		200
6.10	Summary		201
	References		201

Chapter 7 – Risk acceptance criteria   204

7.1	Introduction		204
7.2	Decision-makers and society		206
7.3	Risk perception		208
	7.3.1	Acceptable and tolerable risk	208
	7.3.2	Objective risk	209
	7.3.3	Psychological aspects	209
	7.3.4	Social, cultural and institutional processes	213
	7.3.5	Risk communication	214
	7.3.6	Discussion	215
7.4	Formal decision analysis		216
	7.4.1	Objectives and attributes	216
	7.4.2	Expected value analysis	218
	7.4.3	Expected utility	221
	7.4.4	Other techniques	227

7.5	Regulatory safety goals		227
	7.5.1 Types of regulations		227
	7.5.2 Demonstrating compliance - the 'safety case'		229
	7.5.3 Specification-standard regulations		230
	7.5.4 Quantifiable performance requirements		231
	7.5.5 Quantitative safety targets		232
	7.5.6 Examples – quantitative safety targets		233
	7.5.7 Some issues		237
7.6	Quality assurance and peer review		241
7.7	Summary		243
	References		244
Appendix A – Applications			249
A.1	Introduction		249
A.2	Nuclear power plants		249
A.3	Chemical storage facility		252
A.4	Thermal protection system of the Space Shuttle orbiter		256
A.5	'Calibration' – structural reliability of tension members		259
A.6	Gravity dam		261
	References		266
Index			269

Preface

This book arose from our perception that the time was ripe for an exposition of risk analysis which did not have a close bias towards any specific engineering discipline but was rather broader in its focus. We have moved towards this point of view over a number of years, largely as a result of our own consulting involvement in a number of engineering based risk assessment projects and our research activities. The need to look at the broad picture and to attempt to show how various approaches were related and inter-related was reinforced, rather more recently, during the risk assessment conferences held in Newcastle, Australia during 1993 and 1995, both of which we organized. These were very successful events, with participants from industry and from research organizations. The overriding impression was that there is still some way to go in achieving across-discipline understanding about the nature and evaluation of risk, about the need for and limitations of numerical as well as subjective approaches to risk analysis and for an appraisal of the value and limitations of probabilistic risk assessment as is increasingly being demanded in risk-based regulatory approaches.

There is still a degree of misunderstanding about risk assessment and what it can and cannot achieve, whether the numbers that might be produced have any meaning, and whether there is any point in dealing with numbers at all. Unfortunately, some find the simple mathematics required for probabilistic or quantified risk analysis too complex, or pretend that the numbers, even if they could be produced, have no meaning and cannot be related to real life. We do not subscribe to such a pessimistic view, arguing instead that any system of ranking, implicit or explicit, ultimately involves numbers and that it is better to use the best possible numbers obtained by rational analysis of the system rather than some arbitrarily derived set. Of course, much depends on the care, detail and knowledge that goes into any analysis. Sensible results cannot be expected on the basis of poor understanding of the system being studied, poor data and poor computational techniques.

It is our perception that there are 'islands' in the risk assessment world, with different engineering (and other) disciplines having developed, probably quite naturally and deliberately, their own formalism and associated language. Not always have missionaries gone out to talk to the natives on the other 'islands'. While risk assessments are confined within a given industry or engineering (or other) discipline, this is probably perfectly satisfactory. But increasingly risk assessments are being required across discipline boundaries and it is not always possible simply to move a particular technique across a boundary without expecting some problems. Hence a common understanding is desirable, so that the terms and concepts used can be available to other professionals, and that the techniques are transparent and portable across the disciplines. This is one of the main aims of this book. It has not been an easy task, and has required us to focus on basic principles and to restrain ourselves in places from entering into areas where we might have preferred to say more.

A rather successful writer of modern novels recently said that many novels are greatly improved if the author had simply thrown away the first chapter. We hope this is not an example readers of this book will follow. Although Chapter 1 is not particularly exciting it does attempt to set the scene, and suggests the framework for much of what follows. Readers familiar with risk analysis may well want to skip rather quickly here.

Chapter 2 describes various ways in which hazardous situations might arise, drawing on examples from a wide range of industries. In a sense this chapter sets the scene for the issues many engineers and others face in design, construction and management of projects which could have serious impacts on the lives and well-being of people. Chapter 3 introduces the way risk analysts tend to examine the system with which they have to deal. A number of approaches are outlined, from simple but powerful techniques to much more detailed analyses as might be required for complex systems having major potential impacts should failure occur. This chapter essentially outlines the framework within which the next three chapters are set.

Chapters 4 and 5 provide information about the data and the models that need to be input into detailed system analysis procedures such as those described in Chapter 6. Mechanical, physical, electrical and other properties of system components are required, but also information about the demands which are likely to be placed on the system. Of particular importance here is the joint probability of occurrence of particularly onerous or serious demands. Human actions also play an important part in the reliability of systems and Chapter 5 provides an extensive overview of the ways to consider human error, human reliability and the possible control mechanisms related to human error.

Although there is little mathematics in much of the material prior to Chapter 6, the analysis of systems using probabilistic ideas cannot do without it, and consequently this chapter cannot avoid it. Much of the mathematics has been kept as simple as possible. It has been the aim to focus on essentials. In doing so both traditional analysis of fault and event trees is described, but also their analysis when the data for individual events is known only in a probabilistic sense. This is not a topic usually considered in risk texts, yet it is fundamental to realistic risk analysis, and increasingly being employed in practice. For some types of demand on a system the response of components or elements is not available from data banks and must be estimated from a reliability analysis of the component. This aspect is treated using 'first order second moment' concepts as well as more conventional Monte Carlo simulation.

The approaches which might be used to decide on the acceptability of the risk estimated using the techniques of Chapter 6 are reviewed in Chapter 7. Matters such as risk perception, the psychology of risk and social, cultural and institutional issues are discussed and the influences these may have on decision-making are raised. There is also a description of formal risk-based decision-making approaches. The chapter concludes with a review of regulatory safety goals as decision criteria – an important area of immediate application for the concepts, principles and techniques described in this book.

Areas in which the principles and techniques described in this book may be applicable include: nuclear facilities, chemical and petro-chemical industry facilities, off-shore oil platforms, aerospace facilities, civil engineering projects (dams, flood controls), structural engineering etc. Experimentation with the system, such as is common in electronics engineering (for which components are typically cheap, readily replaceable and for which highly redundant systems are cost-effective) is not an option for these systems.

In short, application of the principles and ideas described herein is particularly suited to major projects which are expensive investments and for which prototypes are not possible or are very expensive or are possibly unreliable (as in some chemical engineering systems). For these types of projects there is a very strong need to 'get it right the first time'.

Our thinking about and understanding of risk analysis issues has benefitted from discussions and interactions with many people over many years. Those concerned will know that we appreciate their help and assistance with very considerable gratitude. We hope that their input is reflected in this book. Like most books, this one is also founded on the work of many, many others, not known to us personally. We simply acknowledge our debt and hope that we have not done them too much injustice. As always we have had to be brief, selective and ruthless to meet our objectives within the freedom available to us. Nevertheless, we hope that we have presented a reasonably balanced view of the methods and techniques available for risk assessment of engineering (and other) systems.

Mark G. Stewart and Robert E. Melchers
Newcastle, Australia

CHAPTER 1

Introduction

1.1 RISK ASSESSMENT

Risk assessment has to do with making decisions. For engineering and similar systems, the decisons are about the viability of a new or proposed system or about the continuing viability of an existing system. 'Systems' might include nuclear power plants, railway systems, bridges, petrochemical installations and many others, including software and similar systems.

Whether a system is viable depends on the requirements that are placed on it. A system must be able to fulfil the requirements for which it was established, it must be economical and it must perform at an acceptable level of safety. A little reflection will reveal that is not usually a simple matter to decide whether these three requirements are achieved. Each carries an element of uncertainty – to what degree of certainty must the system fulfil and continue to fulfil its function? what is the probability of the system not being or becoming uneconomic? how likely is it that it will operate safely and continue to do so? To each question could be added the further question: 'what will happen if ...?' that is, what will be the likely consequences if the functional, the economic and/or the safety requirements are not met?

Each of the questions has three components:

1. requirement to be met,
2. the probability of achieving that requirement, and
3. the consequence(s) if the requirement is not met.

Each of these components forms part of the information needed for rational decision-making. The central topic of this book deals with the second of these components – that of estimating the probability that the requirement(s) placed on the system are satisfied.

The procedures for determining these probabilities have a degree of commonality across many industries and disciplines. In contrast, the definition of the requirements a system must satisfy usually are system-, industry- or technology-specific. Similar remarks apply for the determination of consequences. These two matters are not, therefore, central topics of this book, although they are very important components in decision-making processes.

The way the three components (requirement, probability, consequence) relate to each other for particular systems is illustrated in Chapter 2. It will be seen that probability analysis is only part of the overall decision-making process. The assessment of what causes systems to fail is very important, as is the assessment of possible consequences following failure. These aspects should not be forgotten in the detailed discussions to follow.

Since risk assessment has to do with making decisions, it can be seen as another management tool, capable of being applied in a variety of situations, not necessarily the engineering systems which are the primary focus of this book. Indeed, risk analysis and assessment as a management tool has been codified (e.g. AS/NZS 4369, 1995). In this context, application may be at an individual level, or at the level of groups, organizations or at the level of local, national or international society.

1.2 RISK ASSESSMENT IN DECISION-MAKING

Modern approaches to decision-making have attempted to make the risk aspect more explicit. This has not always been the case. Nor has risk always been interpreted in relation to probability.

It is possible to make decisions based on intuition. Usually this applies for relatively minor issues, or where only one or very few people are involved. Where the consequence might be large (e.g. many people involved, much damage possible etc.) decision-makers are usually under pressure to become more rational in their approach. Difficulties usually arise when the data or information base is inadequate to come to firm alternatives. It is then possible to move to voting systems, based on ignorance, popular appeal, demagogues etc. Alternatively, various attempts might be made to rank or catagorize available information and also attempt to do so for matters about which not a great deal is known. Terms such as 'professional judgement' may then be used, and this can be refined by the use of several experts, such as in a Delphi setting. One step further is to invoke more refined ranking systems, such as 'fuzzy set theory', for which there exists an axiomatic basis, or probability theory, for which there exists a somewhat more universally accepted axiomatic basis. The choice of procedure, and hence of theoretical framework, creates in some a deal of passion. The choice in this book is probability theory, being the most widely accepted, consistent with management decision theoretical approaches (as in Cost-Benefit-Risk Analysis) as well as having widespread support from the mathematical community.

To be sure, the use of probability theory for risk assessment does not dispose of some interesting theoretical problems. In particular there is an expectation on the part of many people that the probabilities estimated for the occurrence of failure or other event for a system can always be compared to the frequency of occurrence of specific events such as thunderstorms, or the risk of death caused by thunderstorms, or by aircraft travel, or by mountain-climbing etc. Such 'observed' frequencies make interesting yardsticks but can be used only against properly calculated estimates for the probabilities associated with system events. In particular, account must be taken of all contributors to risk, not just some of them. A criticism often voiced is that proper account is not always taken of human error. When this is the case it should be clear that the calculated probabilities have only 'notional' or relative value. Such notional probabilities have, however, practical uses and have been used in many applications, such as when alternatives within one facility are being compared and external influences can be ignored, and in the 'calibration' of structural design rules (Melchers, 1987). None of this should be cause for alarm or reason to discount the use of probabilities, provided it is understood what the limitations are with which probabilities are estimated for a system. Further, as should be obvious, adopting alternatives to probability theory does not dispose of the problem.

As will be discussed in more detail in Chapter 7, expected frequencies of occurrence increasingly are being used as the foundations for governmental regulatory requirements for particular systems. In view of the above comments, this means that it is essential for a risk analysis to deal with all factors influencing probabilities of occurrence, including human error. Chapter 5 deals with this important topic at some length.

1.3 DEFINING RISK

1.3.1 Risk versus probability

It is necessary to be more clear about the use of the word 'risk'. There are various ways in which 'risk' is understood by different audiences. For statisticians and many engineers, it is simply another word for probability of occurrence of a defined event, as in 'the risk (i.e. probability of occurrence) of a thunderstorm tonight is 20%'. For most people, such an event conjures up the possibility of damage and it is this more than the event itself which causes possible concern. The insurance industry focuses on this type of definition; for them the term 'risk' is simply the (discounted, maybe proportional) value of the items or money 'at risk'. More and more, however, the accepted definition of the term 'risk' embodies both aspects. Thus 'risk' is defined as the probability of particular consequences, evaluated as 'probability × value of consequences'. In this definition the consequences might be evaluated in dollar terms, or in some other valuation system, such as in terms of human fatalities. It is this definition which will tend to be used mainly herein.

The adopted definition of risk will sometimes cause confusion, since at times one of the other definitions is more in keeping with conventional usage. Thus, it is still common to speak of the 'risk of death', meaning the 'probability' of the already defined event: 'death'. Such redundancy should not cause undue problems, however, since in all cases the context will be clear.

As noted, this book is primarily about the estimation of the probability of occurrence of defined events and how such information may be used. Since in most cases consequence estimation is industry specific, it cannot be treated fruitfully in this book. However it is important to note that the level of detail of consequence estimation employed in a risk analysis could have significant implications for the level of detail used to estimate probabilities. For example, interest only in fatalities, rather than various categories of death and injury and perhaps long-term effects leads to a much more closely focused and less arduous risk analysis and, despite its limitations, may well provide sufficient information for certain types of decisions.

The level of detail used in a risk analysis can be broken down in other ways as well. For example, in the nuclear industry, a probabilistic risk analysis (PRA), also more recently, and perhaps for obvious reasons, termed a probabilistic safety analysis (PSA), can be carried out to three levels. These are:

- Level 1 – analysis of the probability of certain critical states being reached in a nuclear plant (e.g. 'loss of coolant failure');
- Level 2 – in addition also analyses of the consequences of various critical states being reached, with associated probabilities; and
- Level 3 – in which further analysis is done for the probable (adverse) effects on humans, including estimation of the loss of life and when this might occur.

To a large extent this book is concerned with Level 1. It has an interest in the other levels only in so far as estimates of probability of occurrence are required. The analysis of consequences (such as the size of an explosion, the spread of radioactive gases in the air etc.) is a matter for others.

1.3.2 Types of risk

Although there will be little need in this book to distinguish between various types of risk, it might be noted, in passing, that sometimes there is a distinction made between 'individual' risk and 'public' or 'societal' risk, and between 'voluntary' and 'involuntary' risk. Both distinctions have something of the same flavour, although they are different. Societal risks relate to groups of people and might be viewed differently to the risk for the individual. One aspect of this phenomenon is the lower acceptability (rather irrationally) of 100 deaths in a group accident as compared with 100 different accidents each involving one death.

Voluntary risks are those willingly undergone by an individual having control over his (or her) actions. Risks such as those incurred in sky-diving, mountain-climbing, smoking etc. are of this type. However, the normal risks associated with living in society, such as those incurred through crime, aircraft crash on housing, domestic gas explosions etc. are usually considered to be of the involuntary type. More difficult to categorize are the risks associated with activities such as driving a motor vehicle. This could be considered voluntary, but is this correct if the only choice to get to work is by car?

Another categorization which is made is between 'real' and 'perceived' risks. Evidently some people have an unreasonable fear about some potential hazards, such as aircraft travel, and this may present a problem in rational decision-making using risk criteria. There may be cultural aspects bound up with the perception of risks. It is a matter for considerable interest in the psychology literature. For the most part, the distinction can be ignored in the discussions to follow, except that it needs to be raised when setting acceptance criteria – a subject for discussion in Chapter 7. Reference might be made also to the Royal Society (1992), Blockley (1992) and Bayersche Ruck (1993).

In the context of criteria for risk acceptance, there is an increasing tendency to distinguish between 'acceptable' and 'tolerable' risks. A little reflection will make clear the distinction between these two terms, but in brief a risk which is tolerable might not be acceptable in the normal course of events. Again, a further discussion of this is given in Chapter 7.

1.4 RISK-BASED DECISION PROCESS

In performing a risk analysis, a number of steps are basic to the analysis, and independent of the system being considered. Nor is it important whether the analysis is being performed for an assessment of compliance with requirements or as a basis for a management decision. The steps are essentially the same, only some of the later parts of the process will vary. The following is based in part on AS/NZS 4369 (1995) but the principles are well-known. Figure 1.1 illustrates the flow-chart for risk-based decisions.

1.4.1 Context definition

A risk analysis should take place within a well-defined context. This means that the relationship between the risk analyst, the system being

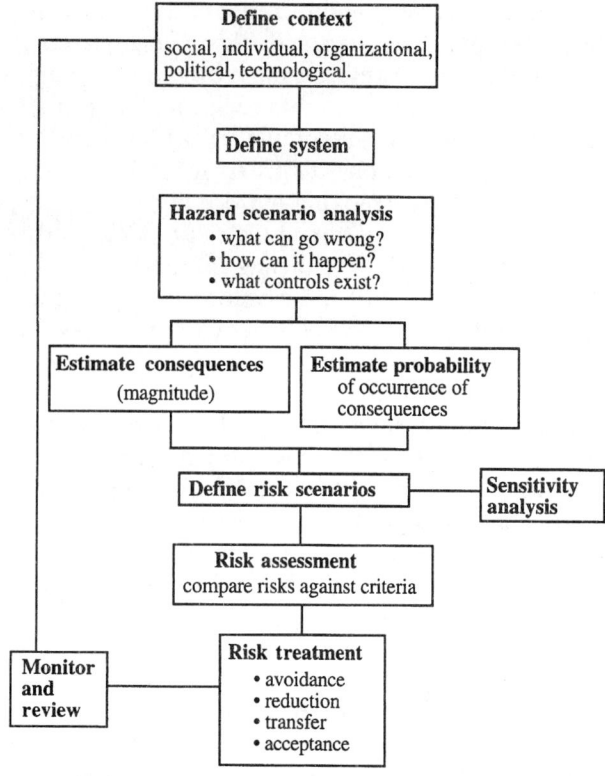

Figure 1.1 Flow chart for risk-based decision-making.
Source: based on AS/NZS 4369: 1995.

examined and the internal and external influences must be known and defined. These include:

1. stake-holders and interested parties, such as society, local communities, individuals, various organizations at local, state and federal level, the organizations directly involved (e.g. the 'client') and groupings within these organizations;
2. matters which might impair the effectiveness of the risk analysis and any recommended action flowing from it;
3. potential influences on the manner in which the risk analysis is performed, such as political, social, cultural, legal and financial aspects.

Organizational issues may also be important. Understanding of the relationship between the system and organizations which bear direct responsibility for it, or which have a direct influence on it, may be important. This will include existing and future goals, objectives and

Risk-based decision process

strategies as these will have an influence on actions that need to be taken (e.g. to ameliorate any adverse effects identified in the risk analysis).

Finally, the risk management context must be defined. The first part of this is the reason(s) why the analysis is being done in the first place. Further, the system to be analysed must be defined, including justification of any parts of the system not to be considered. The comprehensiveness of the analysis and those parties who will bear responsibility for the analysis and the implementation of its outcomes must also be defined.

1.4.2 Criteria definition

Criteria against which the results of the risk analysis are to be assessed are crucial. Various aspects may need to be considered, including cultural, humanitarian, social, legal, financial and technical aspects. For systems in the public domain, national criteria may well exist and be applicable, but for internal, inter-company analyses other criteria are likely to be used. However, even in these cases, it is only rarely that accountability to national or international criteria can be avoided. A more detailed discussion of this aspect is given in Chapter 7.

1.4.3 Hazard identification

The next step is the first of the actual analysis components; namely the defining of the system and its examination to estimate how the system might fail or what might occur under adverse circumstances. There are three steps in the process.

Step 1 Define the structure of the system

In this step, the system usually is decomposed into a number of smaller components and/or subsystems. Together these form a framework for the logical and numerical analysis of the system. They are discussed in more detail in Chapter 3 and in numerical detail in Chapter 6, using data described in Chapters 4 and 5.

Step 2 Identification of hazard scenarios

Identification of what might 'go wrong' with the system or its subsystems is crucial to a risk analysis. This produces the 'hazards' associated with a system. It requires the system to be examined (and understood) in considerable detail. Information from databases, and other past experience, will play an important part in hazard scenario identification. Experiences such as discussed in Chapter 2 and in Appendix A for particular systems will play an important part. Also, various special techniques might be used, such as checklists and brainstorming. For nuclear plants or process plants where a hazard might be considered to have occurred if part of the plant is damaged, the hazard scenarios are sometimes termed 'plant damage states'.

Step 3   Hazard scenario analysis

Once the potential hazards have been identified, it is necessary to identify how such hazards can be realized, that is, how they might come about. It is not sufficient to know the most likely causes – less likely ones may also be important and need to be considered, even if they are later discarded. Moreover, common-cause failures and other linkages between contributing factors can be important and yet understanding of these is often poor. A more detailed discussion is given in Chapter 3.

1.4.4 Risk analysis

This part of the analysis is concerned with determining the probabilities associated with each hazard. To obtain these, the components and sub-sytems of the system are analysed for probability of ocurrence and the various results combined. The essential steps are as follows.

Step 1   Decide on approach and detail

Various techniques can be used for a risk analysis, ranging from qualitative techniques such as ranking, to various quantitative techniques. A further discussion is given in Chapters 3 and 6; the following assumes quantitative approaches.

The level of detail considered in the analysis is in part based in the required outcome and the need for consistency between analysis and modelling detail and output. Evidently, it is desirable to eliminate risks having only minor impact on the outcome at an early time. Preliminary analyses are often used for this purpose. The excluded risks should be recorded, however, to highlight that they were not simply forgotten.

Step 2   Probability of occurrence

In a quantitative analysis the evaluation of risk requires estimation of probabilities (or likelihoods) of occurrence and of the consequences, given the system and its operating conditions. Typically, the probabilities are estimated from a combination of relevant data and subjective judgements, see Chapters 4 and 5. For some situations or sub-systems, the probability of occurrence must be determined (analytically) from information about sub-system capacity and the demands placed on it. This aspect is discussed in Chapters 4 and 6.

Step 3 Consequence prediction

As noted, the prediction of consequences is not usually a matter for risk analysts. Generally input is required from experts in each of the areas for which particular hazards and consequences have been identified. For example, in hazardous gas escapes, topics which require expert advice will include:

(a) the prediction of flow rates,
(b) the manner of spreading of gas and hence the areas at risk,
(c) the probability of ignition,
(d) the likely effect of ignition in terms of heat radiation and/or blast pressure etc., and
(e) the effects, including those on humans.

The expert advice required depends, clearly, on the system under consideration.

1.4.5 Sensitivity analysis

In Chapter 6 it is suggested that imprecision (or 'uncertainty') in data sources and in modelling of the system can be considered using probabilistic approaches to risk analysis. A more traditional approach for estimating the effect of imprecision or uncertainties is to perform a sensitivity analysis. This means that the effect of changes in input variables (data) or in model assumptions is estimated. This provides the so-called 'sensitivity' of the risk estimate. Greater sensitivity suggests that more care and/or effort is required in obtaining data or estimates for the variable concerned. A sensitivity analysis can highlight those parts of the analysis which are of particular importance.

1.4.6 Presentation of results

The results obtained in a risk analysis may well be complex. They need to be presented to potential users in a way offering maximum clarity. A common approach is to group similar hazards and convert these to a common scale, such as lives lost or at risk (i.e. a consequence scale). This can then be related to the probability of occurrence of various such hazards, and might be compared to other hazardous activities or to risk criteria, see Chapter 7.

1.4.7 Risk assessment/criteria

The term 'risk assessment' is usually reserved for the comparison of estimated system risk to criteria of risk 'acceptability'. These are usually established or adopted rather arbitrarily, often using past experience as a guide. Thus the acceptable or tolerable risk of death is often quoted as less than 10^{-6} per year for those not involved in the system under consideration. This figure is based on the expected rate of (premature) death for relatively safe involuntary activities (see Chapter 7). It is important to note that such figures are based on judgements about expectation.

Evidently, comparison to acceptance criteria implies that the risk analysis must produce results which are comparable to the criteria. This means that the analysis must attempt to provide expected probabilities rather than nominal values. Naturally, the consequences must also be comparable to any criteria which may exist.

The risk assessment will produce a list of those hazards making the greatest contribution to system risk. It is these which can be considered for further action if necessary.

1.4.8 Risk treatment

If the estimated risk for the system exceeds either the pre-determined criteria or is in some other way considered to require reduction, there are several options:

Option 1   Risk avoidance, which usually means not proceeding to continue with the system. This is not always a feasible option, but may be the only course of action if the hazards or their probability of occurrence or both are particularly serious.

Option 2   Risk reduction, either through reducing the probability of occurrence of some events, or through reduction in the severity of the consequences, such as downsizing the system, or putting in place control measures.

Option 3   Risk transfer, where insurance or other financial mechanisms can be put into place to share or completely transfer the financial risk to other parties. This is not a feasible option where the primary consequences are not financial.

Option 4   Risk acceptance, even when it exceeds the criteria, but perhaps only for a limited time until other measures can be taken.

In all cases, the proposed course of action requires careful evaluation. Consideration must be given to possible options and to the likely effect of their implementation. This might involve one or more new risk analyses to gauge the effect of changes. In other cases the sensitivity analysis for the original risk analysis may be sufficient to supply the necessary information.

1.4.9 Monitoring and review

Usually a risk analysis presents only a 'snapshot' of the risks associated with a system. Changes to the system, such as its gradual deterioration due to ageing of materials or changes to demands will change the system risk. Also, the effectiveness of control procedures may slacken with time and this, too, will affect the system risk. One way of dealing with such changes is to attempt to predict what the changes might be for some future point in time and then to assess the risk for that time. Naturally, this requires high-quality models to predict change.

An alternative is to monitor the system. In this approach re-evaluations of system demands, system capacities, material strengths and properties etc. are made on the basis of sampling or testing or observation. The risk analysis can then be repeated with the new values for the various parameters and a re-assessment made of the system risk, against acceptance criteria. If the risk analysis has been properly documented, such a revised analysis ought to be relatively straightforward. The modern trend is to record the principal features of the analysis in a suite of computer programs, complete with the various mathematical models used for estimating probabilities and consequences. The result is a risk estimation tool which can supply, with relatively modest effort, an updated risk analysis. In the nuclear industry, this has been termed, appropriately, a 'living' (probabilistic safety) analysis.

1.5 SOME OTHER ISSUES

There are a number of matters not so far discussed which are important for risk analyses. These matters include managerial commitment, documentation and quality assurance, politics and public participation, high consequence–low probability events and legal interpretations of risk. Brief comments about these matters follow.

1.5.1 Managerial commitment

Evidently, a risk analysis will be performed only if there is a perceived need for it. Some analyses will be driven by regulatory requirements, others may arise as a result of managerial policy or need. Whatever the driving force, it should be obvious that the analysis is unlikely to succeed without appropriate managerial commitment. This will include defining policy and objectives and making available sufficient resources for performing the work. It will also require a commitment to communicate the results to those who might be affected. A commitment to ongoing monitoring and review if this is appropriate is also required. Some useful guidelines for implementing a risk management programme within an organization are given in AS/NZS 4369:1995.

1.5.2 Documentation and quality assurance

In the performance of a risk analysis it is essential that a process be instituted for the proper recording of the analysis. The purpose of documentation may be summarized as:

1. to provide transparency of the process and assumptions etc. and thereby to demonstrate that the analysis process was conducted properly;
2. to record critical decisions and assumptions;
3. to provide a record of all data employed, procedures instituted etc.;
4. to provide information for subsequent audits or updating.

Some guidelines for appropriate documentation and quality assurance procedures in simple risk management systems is given in AS/NZS 4369: 1995. Additional comments about quality assurance are given in Chapter 7.

1.5.3 Politics and public participation

The politics of risk assessment, the setting of risk criteria and the use of risk management techniques involves issues such as who decides what the critical risks and consequences are, what the acceptance standards shall be and how much will be disclosed and to whom. Evidently, answers to these matters cannot be given here, nor are they straightforward. Much will depend on the context in which the risk analysis is to be performed, whether in the private domain, or in the public domain and what the nature of the hazard is as well as what the possible consequences of system failure might be.

For public systems, or those which have a potential to have public impact, it is commonly assumed that the public have a right to be properly informed about the risk analysis process, its assumptions and the parameters involved. It is also commonly assumed that the public will have a right to be consulted about appropriate risk acceptance criteria, including discussions about their levels. However, it will be clear from the literature that this has not always been the case in the past, for various reasons, and may not be the case in all cultural or political systems.

Where public participation does occur, it is important that it depends crucially on openness and frankness, as nothing will destroy public confidence more quickly than the perception that 'things are being hidden'. Negotiation, the ability to compromise within acceptable limits and the working towards 'win–win' outcomes are all aspects which need to be considered. However, in the end, it is usually unlikely that everyone can be completely satisfied. The political will to proceed in a desired direction is then crucial in achieving an appropriate outcome. Experience shows that it is seldom logic alone which carries the day.

Largely these matters are outside the scope of the present book. They are signalled here to alert risk analysts to some of the issues which might be involved. Further reading and some useful discussion of these and related issues can be found in the report of the Royal Society (1992), Blockley (1992), Bayersche Ruck (1993) and in Freudenberg (1988).

Part of the discussion which is likely to occur in the public domain is the issue of risk accumulation, that is the build-up of risk as a result of new risks being imposed over existing ones. Evidently, if there are many such new risks, even if they occur over time, the total risk exposure for individuals will increase, at least in principle. For many projects the additional risk may be near to negligible, but this is not always the public perception. For others the issue of risk accumulation is real, more so if it cannot be shown that the risks estimated for existing risk sources have decreased.

1.5.4 High consequence–low probability events

An area of particular difficulty for risk analysis is that where the potential consequences are extremely large or severe yet the probability of these consequences actually occurring is estimated to be extremely low. It is sometimes argued that conventional risk analysis breaks down under this condition, since the risk is now a very large number multiplied by a very small number! More usefully, it should be recognized that it is not the risk analysis methodology which is at fault here, but our ability to use the information obtained from the analysis for decision-making. Where extremely large or severe consequences are predicted, it may be that the decision as to what action to take will hinge mainly on the potential impact of the consequences, irrespective of the risks which are estimated.

Two examples might be useful. The first is the reaction in the United States to the construction of more nuclear power plants following several 'accidents' in the US and the 1986 Chernobyl event in the former Soviet Union. An opposite, less well-known, example is the response to the 1995 Kobe, Japan earthquake, which revealed the difficulty of being adequately prepared for a major, unexpected earthquake. The cost of upgrading all infrastructure in a major city and its environs to cope with a really major but very unlikely natural disaster may be too great for a society to bear, and the best that might be done is simply to resign to a position of 'mopping-up' after an event. In neither case would a conventional risk analysis be particularly useful in making a decision, but at least a risk analysis would reveal the issues.

1.5.5 Legal interpretations and risk

It is sometimes argued that risk-based decision-making leaves engineers, managers and others open to liability when 'the event occurs'. This is a rather simplistic view. It ignores the fact that risk-based design has been used for many years in flood-control, in dam design, in structural engineering generally and in location planning for certain hazardous industries (e.g. the siting of petro-chemical plants). It also ignores the fact that such systems were designed and constructed using 'rules-of-thumb' prior to the use of risk-based decision criteria. All that has happened is that the risk involved in these systems has become more explicit: it was, of course, always there. It should be noted also that risk-based regulatory requirements are legally imposed by societies through their governments and that this will have an effect (although perhaps only in the 'long-run') on the legal system.

The difficulty for some is that courts of law demand that engineers and others must be prudent and take all reasonable measures to ensure the safety of the public and of individuals. It is in the interpretation of these demands that legal argument sometimes overrides what might otherwise have been considered to have been quite prudent actions. But equally there is a tendency to attempt to hide under 'risk' matters which are more properly described as failing to ascribe to a duty of care, including negligence. One could not expect society to accept negligence in performing in a professional manner as covered by a conventional risk analysis.

It is often claimed that the legal profession is perhaps somewhat behind community expectations. Nevertheless, both are moving away from the notion that structures, industrial facilities and the like can be 'perfectly safe'. The difficulty is how to reconcile gross errors, negligence and similar actions or omissions from matters properly considered in a risk analysis and therefore part of the risk accepted by society for a facility. This is not a matter which it appears will be easily or quickly resolved.

For the purposes of this book it is sufficient to point out that there is an ongoing issue regarding the relationship between explicit risk-based decision and the liability of professionals when problems occur. It is suggested that, fundamentally, this is not a new issue, or one which has arisen purely as a result of the introduction of risk-based decision methodologies. It is acknowledged that the issue will not be resolved easily. But it is also pointed out that there are already plenty of antecedents for risk-based decision mechanisms being accepted by society. And in the end that is the essential criterion.

1.6 CONCLUSION

This chapter has given an overview of the concept of risk and its relationship to risk management ideas. A review was given also of the risk-based decision process, of the general role of probability in risk assessment and of risk management. Further, there was a brief overview of some other issues in risk-based decision-making, including the political and legal fields within which risk-based decision-making must operate.

REFERENCES

Bayersche Ruck (ed.) (1993) *Risk Is a Construct*, Knesebeck, Munich.
Blockley, D. (ed.) (1992) *Engineering Safety*, McGraw-Hill Book Co., London.
Freudenberg, W.R. (1988) Perceived risk, real risk: social science and the art of probabilistic risk assessment, *Science*, **242** (7 October), 44–9.
Melchers, R.E. (1987) *Structural Reliability: Analysis and Prediction*, Ellis Horwood, Chichester, England.
Royal Society (1992) *Risk: Analysis, Perception and Management*, Report of the Royal Society Study Group, London.
Standards Australia (1995) *Risk Management*: AS/NZS 4369:1995, Sydney.

CHAPTER 2

Sources of risk

2.1 INTRODUCTION

System failure occurs when the system is unable to fulfil some 'acceptable' (or predetermined) level of performance. Thus system failure may be a release of latent energy (e.g. vapour cloud, explosion, fire, aircraft crash), release of toxic materials into the environment, interruption to the operation of the system, or any other situation where the system does not perform according to its designed function. Often system failure is caused by a combination of failure events or processes, typically by the failure of one or more individual components that are required to function correctly for the successful completion of the system task. These individual components may include equipment failure, human error (i.e. human action that exceeds some level of acceptability), excessive loads etc. In this chapter it will be useful to consider the general factors that contribute to system failure (failure events), and their causes, for a range of engineering systems, such as:

1. nuclear power plants;
2. chemical and oil process and storage plants;
3. shipping;
4. pipelines;
5. structures;
6. offshore platforms;
7. dams;
8. hydrologic systems.
9. aerospace systems;
10. robots; and
11. computer systems.

The above list comprises a diverse range of systems, but 'what large systems have in common counts for more than how they differ in design and intention' (Bell, 1989). As will be seen, the sources of risk often are common to many engineering systems.

Some of the above systems might be classified as sub-systems of other engineering systems. For example, the chemical and oil industry 'system' may include offshore platforms, process and storage plants, pipelines, computer systems and shipping.

Sources of risk for specific engineering systems

This overview presents an important starting point for comprehensive hazard scenario analysis (see Chapter 3) because it will highlight those components of engineering systems which have been shown to have an adverse influence on system performance. Nevertheless, it is important to note that even with the best intentions, advice and extensive experience and databases, it may not be possible to identify all potential failure events or hazards in an engineering system. Some failure events, referred to as 'unforeseen' events or processes, are likely to remain. However, a sound hazard scenario analysis should reduce these to a minimum.

2.2 SOURCES OF RISK FOR SPECIFIC ENGINEERING SYSTEMS

The general causes of system risk identified herein are intended to illustrate only the factors that have contributed to system failure. This information may help to:

1. define the nature of risk to be quantified (i.e. define system failure);
2. identify event(s) or sequence of events that may lead to system failure (e.g. pipe or valve failure, high loads); and
3. identify possible causes of these events (e.g. corrosion, end of design life, fire, earthquake, human error).

The engineering systems to be discussed below are complex ones and it will be possible to describe the causes of risk only in generic terms.

It is beyond the scope of this book to suggest measures to ameliorate the likelihood and/or consequence of system failure. Often such measures comprise a range of industry-specific approaches and may be applicable to only a single failure event. Also, it is likely that measures to improve equipment or human performance may be obtained best by examining the exact cause(s) of the event. For example, operator error may be reduced by better training, less time pressure, ergonomically designed instruments, automated controls, etc. On the other hand, the incidence of pump failure may be reduced by more frequent maintenance or the installation of newer pumps. Moreover, the consequences of pump failure may be reduced by the installation of standby and parallel redundant sub-systems and better training and supervision of operators.

2.2.1 Nuclear power plants

System failure of a nuclear power plant (NPP) is generally defined as a release of radioactive material beyond the boundary of the site (typically taken as one mile from the NPP). These give so-called 'off-site' consequences on the surrounding environment. These may include early or latent health effects, loss of habitability and economic losses.

Once system failure has been defined, it is possible to identify events with the potential to initiate system failure. Techniques such as event trees or fault trees (see Chapter 3) may be used to describe events or sequences of events that may lead to system failure. For example, Figure 2.1 shows a fault tree developed for a risk analysis of a NPP situated in the United States (USNRC, 1989). The fault tree outcomes represent undesirable events that may lead to off-site releases (i.e. Levels II–VII), where Level VII events may be defined as sub-system failures. The initiators shown in Level VIII may be described also by other fault trees which would show the equipment or processes that must fail to cause a sub-system failure. The complexity of the system is clear from the observation that it was necessary to consider 16 000 accident sequences (i.e. different paths through the fault tree) in the USNRC (1989) study. A detailed description of event trees, fault trees and other system representation methods is provided in Chapter 3.

Consider now in detail an important aspect of NPP operations and one for which there is some available data; namely, the performance of valves. Valve failure is defined as the failure of the valve to operate correctly, and this may lead to loss of core cooling; which in turn may result in damage to the core. Data on 1842 safety-related events associated with operating experience with valves in light-water-reactor NPP for the period 1965–78 are available (Scott and Gallaher, 1979). This data was used to describe (1) the equipment in which the valve failure occurred, and (2) the causes of valve failure, as shown in Tables 2.1 and 2.2 respectively. The extensive list of equipment in which valve failure has occurred in the past (see Table 2.1) provides a good indication of the large number of potential accident sequences.

Table 2.2 shows that valve failure may be due to one or more of a variety of causes. Human error appears as the dominant cause and it may occur in the administration, design, fabrication, installation, operation and maintenance of the equipment. The remaining causes generally are due to physical effects such as fatigue, corrosion, leaks, and 'end of design life' being reached. It might be noted also that 92% of valve failures occurred during the testing and operation of the NPP, and that the remaining valve failures occurred during construction or refuelling. The influence of external events such as fires, earthquakes, cyclones, flooding and aircraft crashes also need be considered as potential causes of equipment or process failure. For example, Table 2.3 shows some of the structural and equipment items that may be damaged in an earthquake.

A large number of system failures in non-nuclear electric power plants (e.g. power outages and accidents) are also caused by human error (e.g. Floyd, 1986). Remaining system failures generally are due to equipment failure, lightning or utility disturbances.

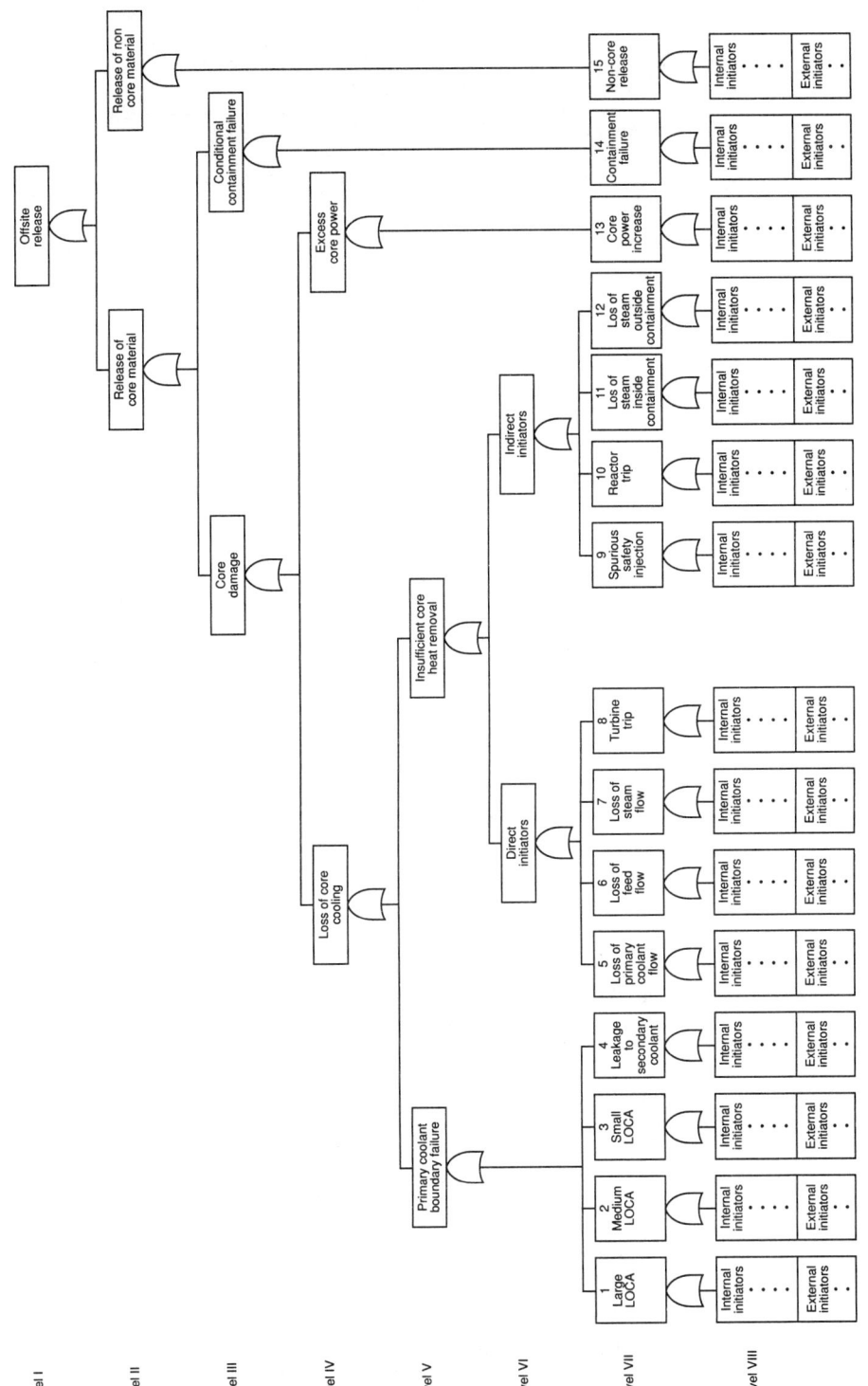

Figure 2.1 Typical fault tree for a nuclear power plant.
Source: USNRC (1989).

Table 2.1 Equipment in which valve failure occurred for boiling water nuclear reactors

Equipment	Frequency (%)
Condensor	2
Control rod drives	2
Filters	1
Generator (diesel)	2
Heat exchangers	1
Instruments	19
Motors	4
Piping	2
Pumps	3
Radiation monitors	2
Relays	3
Seals	6
Sensors	5
Solenoid	7
Turbine	4
Valve operators	16
Valve (check)	7
Others	12

Source: adapted from Scott and Gallaher (1979).

Table 2.2 Cause of valve failure for boiling water nuclear reactors

Cause	Frequency (%)
Physical causes:	(54%)
Age	1
Corrosion	1
Crud	6
Erosion	1
Fatigue	1
Natural failures (e.g. end of design life)	21
Flaw	1
Leak	14
Lubrication	2
Stress	1
Vibration	2
Wear	1
Weather	1
Human causes:	(46%)
Administrative errors	3
Plant design errors	11
Fabrication errors	4
Installation errors	5
Operator errors	6
Maintenance errors	18

Source: adapted from Scott and Gallaher (1979).

Table 2.3 Structures and equipment that may be damaged in an earthquake

Yard equipment and structures:	Fuel oil tank, refuelling water storage tanks, auxiliary feedwater storage tank, primary feedwater storage tank, service water intake structure and pumps, auxiliary turbine (including battery and controls), electrical yard and transformers.
Auxiliary building:	Heat exchangers, pumps, tanks, valves, ventilation equipment, component cooling system surge tank, air conditioning equipment.
Control area:	Ceiling, control panels, switchgear in the main control room, walls, cable trays, switchgear, electrical buses, transformers, inverters, battery rooms, control cabinets, ventilation equipment.
Reactor containment:	Steam generators, accumulators, containment fan cooler units, spray headers, containment isolation valves, electrical penetrations, instrument tubing and transmitters.
Diesel generator rooms:	Diesel generators, diesel control area, vital control centres, day tanks, exhaust stack, fuel storage tanks, fuel transfer pumps.
Penetration areas:	Vital control centres, pumps, valves, heat exchangers, compressors, main steam isolation valves, safety valves.

Source: adapted from Ravindra, *et al.* (1990).

2.2.2 Chemical and oil process and storage plants

The processes in the chemical and oil industries often involve production, storage and transportation of highly hazardous substances. System failure may occur in any one of these processes. System failure may be defined as the occurrence of one or more of the following hazards:

1. fire – ignition of leaked/spilled fluid or gas;
2. explosion – ignition of vapour cloud; and
3. toxic release.

Lees (1980) reports that while fires are the most common form of failure, explosions tend to result in higher losses and fatalities. However, a toxic release has the greatest potential for injury or loss of life.

Both the occurrence and the consequences of system failure may be influenced by a number of factors, including (1) the reactivity and toxicity of the materials, and (2) the sensitivity of the materials to the conditions under which they are processed (i.e. high or low pressures and temperatures). The following facilities are considered to be particularly hazardous (Lees, 1980):

1. air and oxygen plants;
2. ammonia plants;

3. ammonium nitrate plants;
4. olefin plants; and
5. LPG (Liquefied Petroleum Gas) and LNG (Liquefied Natural Gas) installations.

Some chemicals also interact with, or are essential components of, other engineering systems. For example, LPG and LNG are generally used as fuels for a variety of industries; hence a fire or explosion of a LPG or LNG installation (e.g. storage facility) may severely damage nearby engineering systems (e.g. nuclear power or hydrogen plants).

Using the FACTS (Failure and Accidents Technical Information System) incident databank, Koehorst (1989) found that about 6% of the major accidents in the chemical industry occurred during the process of startup/shutdown, about 60% during normal operation and the rest during maintenance. Typical causes of fires and explosions in the chemical industry are shown in Tables 2.4 and 2.5. Evidently, equipment failure (e.g. pumps, piping, instruments) and human error appear to be dominant causes of system failure. The types of equipment that are prone to failure in a refinery are given in Table 2.6. It is likely that human error may occur in the process of planning, design, construction, operation, inspection, maintenance, transportation and storage (e.g. Koehorst, 1989). Further, it appears that management or organizational factors are

Table 2.4 Causes of large losses in the chemical industry

Cause	Frequency (%)
Incomplete knowledge of the properties of a specific chemical	11.2
Incomplete knowledge of the chemical system or process	3.5
Poor design or layout of equipment	20.5
Maintenance failure	31.0
Operator error	6.9

Source: adapted from Doyle (1969).

Table 2.5 Causes of fires and explosions in the chemical industry

Cause	Frequency (%)
Equipment failures	31.1
Inadequate material evaluation	20.2
Operational failures	17.2
Chemical process problems	10.6
Ineffective loss prevention programme	8.0
Material movement problems	4.4
Plant site problems	3.5
Inadequate plant layout and spacing	2.0

Source: adapted from Spiegelman (1969).

Sources of risk for specific engineering systems

Table 2.6 Significant equipment failures in a refinery

Equipment	Frequency (%)
Pumps and compressors	33.9
Utilities	22.3
Furnaces	13.6
Piping	10.7
Towers and reactors	8.8
Exchangers	6.8
Other	3.9

Source: CEP (1970). Reproduced with permission of the American Institute of Chemical Engineers. Copyright ©1970 AIChE. All rights reserved.

partially or fully responsible for the majority of safety related incidents (Robinson, 1987).

2.2.3 Shipping

Systems for the transportation of people and the shipment of large quantities of oils and chemicals have the potential to cause severe consequences if system failure occurs. System failure may include structural failure, fires, explosions, and oil and chemical spills. In open, unrestricted waters the consequences of ship system failure on third parties may not be particularly severe. However, an explosion or release of hazardous/toxic materials along the coastline or in a harbour may cause loss of life, environmental damage, and subsequent economic losses. For example, tankers carrying liquefied nitrogen gas have the explosive potential to devastate a substantial part of a city – it has been observed that 'some tankers resemble floating chemical storage plants with dangerous chemicals being maintained at delicate temperatures and pressures' (Perrow, 1984).

Generally collisions, groundings, structural failure, fires, and explosions are the main initiating events for shipping accidents, see Table 2.7 (Bertrand and Escoffier, 1989). A classification of the causes of collisions is shown in Table 2.8; evidently, violations and/or judgement errors account for about 90% of collisions (Gardenier, 1976). It is therefore not surprising that it has been estimated that approximately 80% of shipping accidents are caused by human error (Gardenier, 1981; Perrow, 1984). The sources of human errors may be identified as either (1) unforced operator errors or (2) forced operator errors. Forced operator errors are produced, for example, by error-likely situations such as long periods without rest, and by the threat of financial fines for missing shipping schedules. Approximately 5–10% of shipping accidents appear to be due to the unanticipated interaction of multiple errors or events (i.e. unforeseen events). Other accidents may be caused by equipment failure

Table 2.7 Initiating events causing tanker accidents

	Frequency (%)	
Initiating event	Loaded tankers	Unloaded tankers
Grounding	37.4	28.7
Collision	27.1	12.9
Structural damage	9.4	4.9
Explosion	7.6	39.6
Leak	5.8	3.0
Fire	5.3	7.9

Source: adapted from Bertrand and Escoffier (1989).

Table 2.8 Cause of shipping collisions in the period 1970–74

Cause of collision	Frequency (%)
Deliberate violations of rules of road	55.6
Judgement errors	50.0
Environment	46.5
Vessel design/waterway design	31.3
Late detection	30.0
Multiple vessels	9.5
Mechanical failures	8.0

Note: Some collisions are due to multiple causes.
Source: adapted from Gardenier (1976).

or natural phenomena such as storms, fog or shifting channels (Perrow, 1984).

2.2.4 Pipelines

A pipeline system failure can result in the spillage of toxic liquids or gases into the environment; this may cause widespread pollution and may constitute a serious health hazard. For oil pipelines, it has been estimated that approximately 73% of failures occur in the line pipe itself and appproximately 27% in auxiliary equipment along the line (e.g. valves, flanges, instrument connections). The same study also categorized the cause of failure as:

1. mechanical failure (material failure and construction defects);
2. operator errors;
3. corrosion;
4. natural hazards; and
5. third-party damage (e.g. ship anchor damaging offshore pipeline).

See Table 2.9 for the relative frequencies of these events (Anderson and Misund, 1983).

Table 2.9 Causes of oil pipeline failures

Cause	Frequency (%)
Corrosion	29.9
Third-party damage	27.2
Operational error	26.6
Mechanical failure	10.7
Natural hazard	5.6

Source: Anderson and Misund (1983).

2.2.5 Structures

Buildings, bridges and other structures consist of interconnected beams, columns, connections, floors, walls and foundations to form a structural system. The system failure of these engineered structures may be defined as either (1) structural collapse (either of isolated structural member or of all or part of the structural system) or (2) functionality or serviceability failure (e.g. excessive movement or deflections, or excessive vibration, such that the structure cannot be used for the functions for which it was designed). Structural system failure can lead to death (or injury) of construction workers and/or the public, repair or damage costs and other economic losses. Generally, the nature of the risk to be quantified is the likelihood of structural collapse (i.e. probability of failure) since this mode of system failure often causes the most catastrophic consequences. For this reason, only the initiating events and causes of structural collapse are considered herein.

For structural engineering purposes, structural collapse is deemed to occur when the loads exceed the structural resistance (or capacity) of the structural element or system. Therefore, high loads or low resistances are the initiating events of structural collapse. Loads generally consist of one or more of the following components: dead, live, snow, wind and earthquake. High dead and live loads are generally influenced by the user; for example, overloading, or collision of heavy vehicle, trains or ships with bridge structure. On the other hand, high wind, flood, snow and earthquake loads occur as natural phenomena outside the control of the user. System risk may be increased also as a result of a low structural resistance. This may arise from natural (and expected) variation in material properties or as a result of undersized or weakened structural elements.

The causes of low structural resistance or high loads (or a combination of the two) may be categorized as either (1) natural or (2) man-made (i.e. human error). The available statistical data suggest that human error is the cause of up to 75% of structural failures (e.g. Matousek and Schneider, 1977). However, in areas subject to extreme natural phenomena such as earthquakes and cyclones it is likely that significantly more than 25% of failures will be due to extreme loads.

Table 2.10 Distribution of errors in buildings and bridges

Planning and design	Construction	Utilization and maintenance	Number of failures surveyed	Source
45	49	6	800	Matousek and Schneider (1977)
53	47	–	277	Fraczek (1979)
77	22	1	120	Walker (1980)
64	31	1	10000	Logeais (1980)
43	32	25	87	Hadipriono (1985) – Buildings
12	23	65	54	Hadipriono (1985) – Bridges
39	40	21	604	Eldukair and Ayyub (1991)

Note: The percentage occurrences may not sum to 100% due to the influence of other errors.

Table 2.10 shows that human error can occur in the processes of planning, design, construction, utilization (including external events) and maintenance of engineered structures. These errors may be committed by architects, engineers, contractors (e.g. construction workers), building inspectors and users of buildings. Care is required in interpreting statistics such as these. For example, users are not necessarily responsible for all errors in the utilization of buildings and bridges. In fact, engineers may be responsible in part or totally if they had not anticipated the influence of external events such as vehicle, train or ship impact loads or overloading by people (e.g. people crowding onto a balcony during a parade). In general, it would appear that engineers and contractors ultimately are responsible for most structural failures, as indicated in Table 2.11. It lists the ten primary causes of structural failure as obtained from a study of 604 cases in the United States during the period 1975–86 (Eldukair and Ayyub, 1991). Table 2.11 shows that poor construction procedures are the main cause of structural failures. Note that approximately 7% of structural failures were caused by 'unforeseen events': it is highly unlikely that such events could be incorporated into a risk analysis.

Where there is doubt about being able to identify events leading to structural (or other) failure, every effort should be made to minimize the consequences of possible failures. For example, structures may be designed so that the loss of a single structural component (e.g. loss of a bridge pier due to a shipping collision) does not lead to the collapse of the entire structure or suspension bridges may have to be closed during times of high winds to prevent vehicles blowing over and damaging hanger members (Shiraishi and Cranston, 1992).

Generally similar arguments apply to the structural aspects of aircraft, trains, ships, motor vehicles, aerospace vehicles, industrial plants, and offshore platforms since these contain structural sub-systems and components generally similar in character to those used in buildings and

Sources of risk for specific engineering systems

Table 2.11 Primary causes of structural failure

Primary cause	Frequency (%)
Inadequate load behaviour	45.2
Inadequate connection elements	47.0
Reliance on construction accuracy	1.8
Errors in design calculations	2.5
Unclear contract information	23.5
Contravention of instructions	21.8
Complexity of project system	1.2
Poor construction procedures	54.3
Unforeseeable events	7.1

Source: adapted from Eldukair and Ayyub (1991).

bridges. It follows that the initiating events and their causes as described in this section are also applicable to the structural design, construction and use of these systems.

2.2.6 Offshore platforms

System failure for offshore oils and gas platforms is generally defined to occur when there is either (1) a structural failure of one or more subsystems (usually the foundations, jacket or deck); or (2) a severe operational accident which causes system damage, or loss of production or loss of life.

Initiating events that have caused accidents in fixed and mobile offshore platforms during the period 1955–90 are shown in Table 2.12 (Bertrand and Escoffier, 1991). The initiating events are similar in origin to those experienced in the process industries and structural systems (sections 2.2.2 and 2.2.5). Only a small proportion of accidents were

Table 2.12 Initiating events causing severe offshore platform accidents

	Frequency (%)			
Initiating event	Fixed platforms	Jack-up rigs	Submersible rigs	Semisubmersible rigs
Blowout	34	23	50	28
Fire/explosion	25	6	14	6
Collision	9	5	3	11
Capsizing	8	9	10	3
Structural damage	8	32	7	13
Drifting, grounding	–	8	3	20
Weather, flooding	3	6	10	9
Other	8	11	3	10

Source: adapted from Bertrand and Escoffier (1991).

caused by environmental or natural hazards (e.g. extreme wave or wind loading). Almost all of the remaining initiating events can be traced back to a sequence of human errors (Bea, 1989). Paté-Cornell (1989) has categorized these errors as either procedural or organizational. Procedural errors arise when an operator fails to perform a specific task in a procedure of operations, while organizational errors may result from poor organizational or management decisions. These errors may occur in the design, construction and operation of offshore platforms.

2.2.7 Dams

For dams the failure which is of most interest is rupture of the wall since the resulting floodwave (i.e. the sudden release of water) is likely to cause severe loss of life and a large amount of property damage.

As with the systems discussed above, dam failure is generally the result of the 'loads' exceeding the 'resistance' of the dam. This may be interpreted in terms of loads such as due to earthquakes, large floods or the failure of an upstream dam. The resistance of the dam may be reduced by latent design/construction defects or natural variation in concrete/soil properties. Initiating events may be referred to as external and internal events respectively (Bowles, 1987). The relative occurrence of these events is shown in Table 2.13, for dams built after 1945 (Blind, 1983). It is observed that the dominant initiating events are overtopping (scouring of unprotected dam foundations due to large floods or failure to open gates) and failure of the foundations (e.g. pore pressure, seepage, settlement). Serafim (1981) suggested that most dam failures can be attributed to human error, and very few are a direct result of natural phenomena (i.e. those which are not known about or which are impossible to control). This is in accordance with data given by Loss and Kennett (1987) who have shown that the majority of dam failures are due to human error, see Table 2.14. For example, overtopping may be caused by poor assumptions in determining the design flood or failure to consider the effects of earthquakes.

Table 2.13 Initiating events for dam failure

Initiating event	Frequency (%)
Failure of foundation	36
Overtopping	33
Cracks in dams	7
Slides (banks or dam slopes)	5
Incorrect calculations	1
Unknown reasons	18

Source: adapted from Blind (1983).

Table 2.14 Cause of dam failure

Cause	Frequency (%)
Design error	23
Poor assumptions	12
Poor construction	12
Poor inspection	12
Management/communication	7
Practice error	7
Other	26

Source: adapted from Loss and Kennett (1987).

2.2.8 Hydrological systems

Hydrological systems such as water supply, flood control, waste management, and water quality systems are concerned with water and its influence on people and the environment. These systems may also contain many subsystems; for example, a supply system typically would contain treatment plants, storage reservoirs, pumping stations, transmission and distribution pipelines (Shamir, 1987). For most hydrologic systems, system failure can be defined to occur when the 'load' (or 'demand') exceeds the 'resistance' (or 'capacity') of the system (Duckstein et al., 1987). In this sense loads or demands may include flood volume, pollutant loadings (e.g. chemical hazards), and water demand (e.g. for drinking or irrigation). On the other hand, reservoir flood storage, levee height, and cleaning and supply capacities are typical resistances or capacities for these systems. Unfortunately, there appears to be relatively little data available about the causes for low resistances or capacities and/or high loads or demands for hydrologic systems.

2.2.9 Aerospace systems

Aircraft accidents are considered to arise as a direct result of one or more of the following hazards acting on the aircraft system:

1. collision with the ground, water, buildings or other aircraft;
2. fire (and smoke); and
3. loss of structural integrity (e.g. loss of cargo door, engine falls off).

It is also possible that an aircraft crash may be an initiating event for the failure of other hazardous systems such as nuclear power plants, chemical process plants, buildings and dams.

Typical events that may have been involved in the occurrence of aircraft accidents are given in Table 2.15, for passenger jet aircraft in the period 1969–76 (Lloyd and Tye, 1982). The causes of these events may be broadly categorized as:

Table 2.15 Events leading to passenger jet aircraft accidents

	Frequency (%)	
Event	Fatal accidents	All accidents
Predominantly airworthiness		
Airframe structural failure	1.6	7.1
Fire (cabin, toilet, etc.)	3.2	2.4
Fire (landing gear failure)	1.6	6.8
Fire (engine failure)	7.9	19.6
Landing gear failure	0.0	4.4
Flying control systems failure	11.1	4.7
Predominantly operational		
Striking high ground	22.2	4.7
Undershoot runway	36.6	15.1
Overshoot runway	6.3	9.5
Running off runway	0.0	7.8
Heavy landing	0.0	5.4
Weather	9.5	6.1
Bird strikes	0.0	6.4

Source: adapted from Lloyd and Tye (1982).

1. single or multiple material/equipment failures;
2. human error; and
3. weather or environmental conditions (e.g. ice, snow, ingestion of birds into engine).

Numerous studies have shown that human error is responsible for nearly all aircraft accidents (e.g. Thurston, 1980; Lloyd and Tye, 1982; Gloag, 1991), with errors typically committed by:

1. design and manufacturing staff;
2. maintenance staff;
3. pilots and other flight crew;
4. cabin crew, groundhandlers and passengers; and
5. air traffic controllers.

Table 2.15 shows that most accidents arose from operational events; that is, during landing and takeoff, when the flight crew is considered to be in control of the aircraft. Thus, it is not surprising that most studies show that flight crew error causes most system failures (at least 50% of all) (e.g. Nagel, 1988). However, poor cockpit design or inadequate operating procedures are known to contribute to the occurrence of flight crew error (Lloyd and Tye, 1982). Air traffic control errors appear to cause only approximately 1% of all aircraft accidents; these accidents are generally mid-air collisions (Weiner, 1980). Interestingly, it would appear that few if any major aircraft accident can be attributed to radar or computer failure in an air traffic control system (Driver, 1979).

There appear to be few data in existence on the sources of risk for space vehicles and satellites. Mainly, this is due to (1) the limited operational experience of most space systems (e.g. approximately 150 manned space flights up to 1995), and (2) these systems are often developed and operated by those with military or commercial interests and hence their operating experience is often treated as 'classified' information. Nonetheless, it is reasonable to assume that sources of risk for the design, production and operation of space vehicles and satellites are not dissimilar to those described above for aircraft systems. Additional sources of risk, such as space debris (e.g. meteorites and man-made space junk) and solar radiation (mainly due to solar flares) also need to be considered. Further, it is important to note that the development of aircraft often includes approximately 1000 test flights, and every production aircraft is given at least two test flights before final delivery. This is not the case for space vehicles and satellites as the high cost of producing them (and the fact that most space vehicles are expendable) tends to limit severely the number of test flights or trials. Further, severe weight limitations (made in order to maximize payloads) restrict the number of redundant and parallel systems that can be designed into aerospace systems; thus lowering their safety margins (Ashford and Collins, 1990). For these reasons, it is more likely that these systems will be particularly sensitive to equipment failure, environmental conditions, human error and unforeseen events.

2.2.10 Industrial robots

System failure for industrial robots may be defined as (Khodabandehloo et al., 1984a):

1. undesirable robot movement in playback mode;
2. undesirable robot movement in teach mode;
3. arm runaway when switching on;
4. no emergency stop action when demanded; and
5. arm 'creep' or degradation of repeatability.

System failures are generally caused by the occurrence of one or more of the following initiating events (Khodabandehloo et al., 1984b):

1. random component failure (mechanical, electronic, hydraulic and pneumatic);
2. systematic hardware faults (e.g. design deficiency);
3. computer software faults; or
4. human error.

Random component failures may be caused by radiation, dust, fumes, corrosion, fatigue, and electrical interference (Dhillon, 1988). However, it has been observed by Khodabandehloo et al. (1984b) that human error is the cause of a 'large number of failures'.

2.2.11 Computer systems

Increasingly, computers are used to design, monitor and control complex engineering systems. Some of the systems may be hazardous, while others will be time critical. Examples include air-traffic control, airline reservation and telecommunication systems. Their failure may lead to death or injury, economic losses or environmental damage.

It is perhaps not surprising that computer system failures are the cause of many operational failures in space, military and aerospace systems (Leveson, 1986) as well as other systems. For example, software errors were discovered in a program used to design structural supports for pipes and valves in the cooling system of nuclear reactors (Neumann, 1979). Experience shows that in many cases hardware and/or software failures (e.g. run-time errors) tend to be the initiating events that lead to system failure. These initiating events may be caused by (Leveson, 1984):

1. hardware component failures;
2. interfacing (i.e. communicating) problems between system components;
3. human error in operation or maintenance;
4. environmental stress; or
5. software errors.

It could be argued that since computer software systems do not depend much on hardware (and its failure) human error would appear to be the major cause of computer system failure (Kletz, 1993).

2.3 UNFORESEEN EVENTS

The causes and consequences (damage, health and environmental effects) of some physical phenomena are not necessarily well understood; for example, steam explosion, formation and distribution of hydrogen during a core-melt accident or toxic vapour cloud release (Rodder and Geiser, 1977; Whittmore, 1983). It is therefore possible that unexpected and unforeseen events may occur because of unknown phenomena. However, it is reasonable to assume that appropriate research efforts will lead to increased understanding of unknown phenomena until the phenomena eventually becomes 'known phenomena'. In other words, the proportion of system failures due to unknown phenomena will decrease over time. This process has been illustrated for structural engineering systems in Figure 2.2. Shiraishi and Furuta (1989) observe that system failure will still occur if the causes and consequences of known phenomena are not known to the appropriate authorities (i.e. becomes general engineering knowledge). This new knowledge may be incorporated into codes of practice, incident databanks, checklists etc. Reference

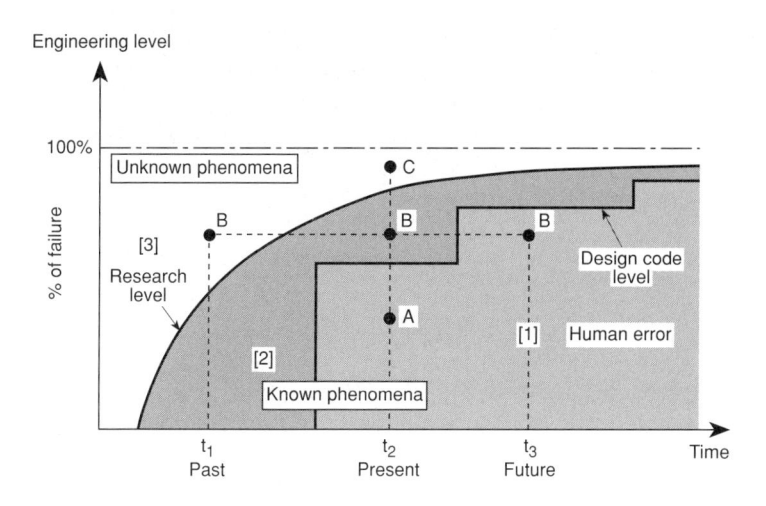

Figure 2.2 Causes of system failure.
Source: Shiraishi and Furuta (1989).

to Figure 2.2 shows that human error is then increasingly the primary cause of system failures if the phenomena that contributed to system failure is known to all.

Changes in knowledge and/or information may also lead to the identification of events that were previously not foreseen. This new knowledge may be obtained from operating experience (e.g. accidents or near-misses may reveal unforeseen accident sequences), measurements of natural phenomena (e.g. wave forces) over a longer period of time, and theoretical or experimental research. New knowledge is most likely to occur during the development and application of new engineering systems. This new information may invalidate prior or current risk analyses. For example, the availability of more rainfall data and improved statistical analyses have lead to higher estimates of the design probable maximum precipitation (PMP). Therefore, it has then been found that many dam spillways (worldwide) are inadequate (Wellington, 1988). Further, new toxicology studies may indicate that the potential for loss of life may have increased; it may then be necessary to update or revise previously conducted risk analyses.

The occurrence of unforeseen events may be minimized by (1) learning from past experience (i.e. from success, failure, near misses) and (2) communicating information from past experience to individuals and organizations that need such information. However, it may not always be possible to communicate this information between individuals, organizations and the engineering community in general because of the poor organizational structures of some engineering systems. For example, information may be buried among other material, information may be distributed among several organizations, or there may be no place for information in existing categories or incident databanks (Turner, 1978). Furthermore, it is recognized that the potential for litigation often restricts the disclosure of some information (e.g. performance problems). Moreover, it is not always possible to extrapolate lessons learnt from past experience to the development of new technologies (e.g. Petroski, 1985).

2.4 SYSTEM COMPLEXITY

Some engineering systems are likely to contain 'complex interactions' between parts, units and subsystems. Complex interactions is defined by Perrow (1984) as 'those of unfamiliar sequences, or unplanned and unexpected sequences, and either not visible or not immediately comprehensible'. Thus, it is not always possible to predict or foresee the behaviour of a complex system. Complex interactions are most likely in engineering systems containing the following characteristics (Perrow, 1984):

1. common-mode or common cause connections (i.e. failure of a component may lead to multiple effects);
2. interconnected sub-systems;
3. feedback loops;
4. indirect information;
5. multiple and interacting controls;
6. limited understanding;
7. limited possibility of isolating failed components; and
8. limited substitution of supplied materials.

These characteristics are generally found in nuclear and chemical process plants, aerospace systems, offshore platforms, and engineered structures. Approximately 10% of all components in these engineering systems produce complex interactions; not an insignificant proportion. However, only 1% of all components in other engineering systems (e.g. dams, rail and marine transport, manufacturing, and assembly line production) produce complex interactions (Perrow, 1984).

The accident at the Three Mile Island nuclear plant in 1979 is a typical example of unforeseen events caused (in part) by complex interactions. For this accident 'the combination of design failures, equipment failures, and operator error produced interactions whose consequences exceeded or were different from those of any individual failure'. Not surprisingly, the operators testified 'that they were bewildered by the event' and 'that the contradictory readings were mysterious to the operators'. These observations led Perrow (1982) to conclude that the 'accident was unexpected, incomprehensible, uncontrollable, and unavoidable'. However, this conclusion may be somewhat pessimistic. Garrick (1992) suggests that the accident sequence did fit (although not exactly) into a sequence identified by existing risk analyses. Further, the findings of two major commissions recommended greater use of quantitative/probabilistic risk analyses.

2.5 RECOVERY FROM FAILURE

The ability to recover from the failure of a part, a unit or a sub-system is an important consideration in the estimation of system risk. A successful recovery may prevent, reduce or minimize the consequences of system failure. As a result, many systems have accident mitigation systems (hardware or software or both) and emergency response procedures (e.g. Kabanov *et al.*, 1993). Not surprisingly, the ability to recover from a failure is dependent upon the characteristics of the system. Systems may be classified as either (1) tightly coupled or (2) loosely coupled systems (see Table 2.16). The ability of systems to recover from failures is related to these categories (Perrow, 1984)

Tightly coupled systems require redundancies, parallel systems, safety devices (e.g. emergency pumps) and safety features (e.g. firewalls) that must be designed into the system. In these systems there is generally little scope for expedient or spur-of-the-moment recovery actions if an unexpected event occurs. On the other hand, with loosely coupled systems there is more time/opportunities to initiate recovery actions when an emergency situation develops. Evidently, the identification of sources of risk is particularly important for tightly coupled systems because it is more difficult for these systems to recover from part, unit or subsystem failures. However, this observation does not necessarily imply that loosely coupled systems are inherently safer than tightly coupled systems.

Table 2.16 Characteristics of tight and loose coupling systems

Tight coupling	*Loose coupling*
Delays in processing not possible	Processing delays possible
Invariant sequences	Order of sequence can be changed
Only one method to achieve goal	Alternative methods available
Little slack possible in supplies, equipment, personnel	Slack in resources possible
Buffers and redundancies are designed-in, deliberate	Buffers and redundancies fortuitously available
Substitutions of supplies, equipment, personnel limited and designed-in	Substitutions fortuitously available

Source: adapted from Perrow (1984).

2.6 SUMMARY

For most engineering systems it is observed that the risks which might be associated with them have somewhat similar sources. These may be categorized as:

1. natural phenomena (e.g. lightning, cyclones, earthquakes, snow);
2. external (e.g. aircraft crash, blast waves, terrorism, sabotage);
3. technical failure (e.g. corrosion, fatigue, insufficient knowledge); and
4. human error (e.g. operator error, poor maintenance, poor management).

From the discussion herein it appears that human error is a major cause of system failure. This is not particularly surprising since engineering systems may also be referred to as 'sociotechnical systems' where the 'technical system is embedded within a social system' (Turner, 1992). A social system is concerned with human factors and its influence on the organization and management of the system. Clearly, it is important that the possible effects of human error be incorporated into risk analyses of engineering systems. Human errors that are particularly relevant to engineering systems include:

1. slips or lapses: failure to achieve a set goal (i.e. unconscious action)
2. mistakes: selecting an inappropriate goal (i.e. deliberate or conscious action)
3. organizational errors: errors that result from poor organizational/management decisions
4. latent errors: errors become evident after a long period of time
5. violations: deliberate violation of rules or accepted practices
6. unforeseen errors: trivial errors cause unforeseen circumstances

A more detailed discussion of these and other errors (and their quantification) is given in Chapter 5.

For convenience, it has been assumed in this chapter that system failure generally occurs as a result of an accident or an abnormal event. In some cases, however, it is possible that unforeseen consequences from the normal or routine operation of an engineered system may constitute a type of system failure. For example, emissions from a waste incinerator may have been deemed safe by regulatory authorities; yet these emissions may eventually result in latent health problems in the surrounding population.

REFERENCES

Anderson, T. and Misund, A. (1983) Pipeline reliability: an investigation of pipeline failure characteristics and analysis of pipeline failure rates for submarine and cross-country pipelines, *Journal of Petroleum Technology*, **35**(4), 709–17.

Ashford, D. and Collins, P. (1990) *Your Spaceflight Manual*, Headline Book Publishing, London.

Bea, R.G. (1989) *Human and Organizational Error in Reliability of Coastal and Offshore Platforms*, Civil College Eminent Overseas Speaker Programme, Institution of Engineers, Australia.

Bell,.T.E. (1989) Managing risk in large complex systems, *IEEE Spectrum*, June, 21–52.

Bertrand, A. and Escoffier, L. (1989), IFP databanks on offshore accidents. In V. Colombari (ed.), *Reliability Data Collection and Use in Risk and Availability Assessment*, Springer-Verlag, Berlin, 115–28.

Bertrand, A. and Escoffier, L. (1991) Offshore database shows decline in rig accidents, *Oil and Gas Journal*, **89**, Sept. 16, 72–8.

Blind, H. (1983) The safety of dams, *Water Power and Dam Construction*, **35**, 17–21.

Bowles, D.S. (1987) A comparison of methods for risk assessment of dams. In L. Duckstein and E.J. Plate (eds), *Engineering Reliability and Risk in Water Resources*, Martinus Nijhoff Publishers, Germany, 147–73.

CEP (1970) How reliable is vendor's equipment, *Chemical Engineering Progress*, **66**(10), 29.

Dhillon, B.S. (1988) *Mechanical Reliability: Theory, Models and Applications*, American Institute of Aeronautics and Astronautics Inc., Washington, DC.

Doyle, W.H. (1969) Industrial explosions and insurance, *Loss Prevention*, **3**, 11.

Driver, E.T. (1979) Statement before Subcommittee on Aviation of the Committee of Public Works and Transportation, Washington, 11 December (see also Weiner, 1980).

Duckstein, L., Plate, E. and Benedini, M. (1987) Water engineering reliability and risk: a system framework. In L. Duckstein and E.J. Plate (eds), *Engineering Reliability and Risk in Water Resources*, NATO ASI Series, Martinus Nijhoff Publishers, Dordrecht, 1–20.

Eldukair, Z.A. and Ayyub, B.M. (1991) Analysis of recent U.S. structural and construction failures, *Journal of Performance of Constructed Facilities*, ASCE, **5**(1), 57–73.

Floyd, H.L. (1986) Reducing human errors in industrial electric power system operation, Part I – Improving system reliability, *IEEE Transactions on Industry Applications*, **22**(3), 420–4.

Fraczek, J. (1979) ACI survey of concrete structure errors, *Concrete International*, December, 14–20.

Gardenier, J.S. (1976) Towards a science of marine safety, *Symposium on Marine Traffic Safety*, The Hague, Netherlands.

Gardenier, J.S. (1981) Ship navigational failure detection and diagnosis. In J. Rasmussen and W.B. Rouse (eds), *Human Detection and Diagnosis of System Failures*, Plenum Press, New York, 49–74.

Garrick, B.J. (1992) Risk management in the nuclear power industry. In D. Blockley (ed.), *Engineering Safety*, McGraw Hill, UK, 313–46.

Gloag, D. (1991) Air crashes and human error, *British Medical Journal*, **302**(6776), 550.

Hadipriono, F.C. (1985) Analysis of events in recent structural failures, *Journal of Structural Engineering*, ASCE, **111**(7), 1468–81.

Kabanov, L., Jankowski, M. and Mauersberger, H. (1993) The IAEA Accident Management Programme, *Nuclear Engineering and Design*, **139**, 245–51.

Khodabandehloo, K., Duggan, F. and Husband, T.M. (1984a) Reliability of industrial robots: a safety viewpoint. In T.E. Brock (ed.), *Proceedings of the 7th British Robot Association Annual Conference*, British Robot Association and North-Holland Publishers, 233–42.

Khodabandehloo, K., Duggan, F. and Husband, T.M. (1984b) Reliability assessment of industrial robots. In N. Martensson (ed.), *Proceedings of the 14th International Symposium on Industrial Robots and 7th International Conference on Industrial Robot Technology*, IFS and North-Holland Publishers, 209–20.

Kletz, T.A (1993) Computer control – living with human error, *Reliability Engineering and System Safety*, **39**, 257–61.

Koehorst, L.J.B. (1989) An analysis of accidents with casualties in the chemical industry based on historical facts. In V. Colombari (ed.), *Reliability Data Collection and Use in Risk and Availability Assessment*, Springer-Verlag, Berlin, 601–20.

Lees, F.P. (1980) *Loss Prevention in the Process Industries*, Vols 1 and 2, Butterworths, London.

Leveson, N.G. (1984) Software safety in computer-controlled systems, *IEEE Computer*, Feb., 48–55.

Leveson, N.G. (1986) Software safety: why, what, and how, *ACM Computing Surveys*, **18**(2), 125–63.

Lloyd, E. and Tye, W. (1982) *Systematic Safety*, Civil Aviation Authority, Cheltenham, England.

Logeais, L. (1980) *Statistical Study of Building Structures Behaviour*, IABSE Congress, 125–9.

Loss, J. and Kennett, E. (1987) *Identification of Performance Failures in Large Structures and Building*, School of Architecture and Architecture and Engineering Performance Information Center, University of Maryland.

Matousek, M. and Schneider, J. (1977) *Untersuchungen zur Struktur des Sicherheitsproblems bei Bauwerken*, Report No. 59, Institute of Structural Engineering, Swiss Federal Institute of Technology, Zurich, 1977. (See also Hauser, R. (1979) Lessons from European failures, *Concrete International*, 21–5.)

Nagel, D.C. (1988) Human error in aviation operations. In E.L. Weiner and D.C. Nagel (eds), *Human Factors in Aviation*, Academic Press, San Diego, 263–303.

Neumann, P.G. (1979) Letter from the editor, *ACM Software Engineering Notes*, **4**, 2.

Paté-Cornell, M.E (1989) Organizational control of system reliability – a probabilistic approach with application to the design offshore platforms, *Control-Theory and Advanced Technology*, **5**(4), 549–68.

Perrow, C. (1982) The President's Commission and the normal accident. In D.L. Sills, C.P. Wolf and V.B. Shelanski (eds), *Accident at Three Mile Island: The Human Dimensions*, Westview Press, Colorado, 173–84.

Perrow, C. (1984) *Normal Accidents: Living With High Risk Technologies*, Basic Books, New York.

Petroski, H. (1985) *To Engineer is Human – The Role of Failure in Successful Design*, St Martin's Press, New York.

Ravindra, M.K., Bohn, M.P., Moore, D.L., and Murray, R.C. (1990) Recent PRA applications, *Nuclear Engineering and Design*, **123**, 155–66.

Robinson, B.J. (1987) A three year survey of accidents and dangerous occurrences in the UK chemical industry, *World Conference Chemical Accidents*, Rome, 33–6.

Rodder, P. and Geiser, H. (1977) Formation of hydrogen core during core melt

accidents in nuclear power plants with light water reactor, *Kerntechnik*, **19**, 11.
Scott, R.L. and Gallaher, R.B. (1979) *Operating Experience with Valves in Light-Water-Reactor Nuclear Power Plants for the Period 1965–1978*, Report No. NUREG/CR-0848, US Nuclear Regulatory Commission, Washington, DC.
Serafim, J.L. (1981) Safety of dams judged from failures, *Water Power and Dam Construction*, **33**, 32–5.
Shamir, U. (1987) Reliability of water supply systems. In L. Duckstein and E.J. Plate (eds), *Engineering Reliability and Risk in Water Resources*, NATO ASI Series, Martinus Nijhoff Publishers, Dordrecht, 233–48.
Shiraishi, N. and Furuta, H. (1989) Evaluation of lifetime risk of structures – recent advances of structural reliability in Japan. In A.H-S. Ang, M. Shinozuka and G.I. Schueller (eds), *Structural Safety and Reliability*, ASCE, **3**, 1903–10.
Shiraishi, N. and Cranston, W.B. (1992) Bridge safety. In D. Blockley (ed.), *Engineering Safety*, McGraw Hill, UK, 292–312.
Spiegelman, A. (1969) Risk evaluation of chemical plants, *Loss Prevention*, **3**, 1.
Thurston, D.B. (1980) *Design for Safety*, McGraw-Hill, New York.
Turner, B.A. (1978) *Man-made Disasters*, Wykeham Publications, London.
Turner, B.A. (1992) The sociology of safety. In D. Blockley (ed.), *Engineering Safety*, McGraw Hill, UK, 186–201.
USNRC (1989) *Severe Accident Risks: An Assessment for Five Nuclear Power Plants*, NUREG-1150, US Nuclear Regulatory Commission, Washington, DC.
Walker, A.C. (1980) Study and analysis of the first 120 failure cases, *Symposium on Structural Failures in Buildings*, Institution of Structural Engineers, London, 15–40.
Weiner, E.L. (1980) Midair collisions: the accidents, the systems, and the realpolitik, *Human Factors*, **22**(5), 521–33.
Wellington, N.B. (1988) Dam safety and risk assessment procedures for hydrologic adequacy reviews, *Australian Civil Engineering Transactions*, **CE30**(5), 318–26.
Whittmore, A.S. (1983) Facts and values in risk analysis for environmental toxicants, *Risk Analysis*, **3**(1), 23–33.

CHAPTER 3

Modelling of systems

3.1 INTRODUCTION

The present chapter deals with the modelling of systems. This is a necessary precursor to the numerical quantification of system performance and hence for system risk estimation. As discussed in the previous chapter, there are many sources for and causes of system failure. Assessment of the risk associated with any particular system therefore requires identification of all the ways in which the system can fail. It is then possible to estimate the likelihood of occurrence of such failure modes and what might be the possible outcomes of such failure.

Since the 'system' being considered is often part of a larger system, system failure assessment is a conditional assessment. Typically the output from the failure of one (sub-)system will form part of the input and assessment for the larger system. It is necessary, therefore to understand the inter-relationship(s) between the overall system and its system elements such as the sub-systems and the components of those sub-systems. It is the modelling of this inter-relationship as a first step towards subsequent numerical analysis which is of primary interest in this chapter. Naturally, the identification and representation of the ways in which the system might fail is therefore also of interest.

The modelling of the system may be dependent on the specific characteristics of the engineering system that is being considered. For this reason it should not be surprising that a variety of techniques have been developed in different industries for the modelling of systems. Even though the nomenclature may vary, many of the techniques have similar features, and it will be the underlying unity of ideas which will be stressed in this chapter.

The present chapter will focus primarily on the following components of system modelling:

1. definition(s) of system and system failure;
2. techniques for identification of sources of risk (PHA, FMEA, FMECA, HAZOP, Incident Databanks);
3. techniques for system representation (fault trees, event trees/decision trees).

Introduction

When a system is unable to fulfil some 'acceptable' (predetermined) level of performance, system failure may be considered to have occurred. It follows that the first step in a risk analysis is to define what is meant by 'system failure'. In other words, the nature of the hazard or the mode of system failure must be defined. The next step is to identify individual system elements (e.g. components) that are prone to failure, the potential cause(s) of failure for these elements and the effect of system element performance on the operation of the overall system. It is these failure events which identify the sources of risk for the system as a whole. As already indicated in Chapter 2, such events may include equipment failure, human error, excessive loads etc. This information is then used to develop a system representation, in which logic diagrams (fault trees, event trees) are used to represent the sequences and combinations of events or processes that may lead to failure of the system. Attention also needs to be given to dependency within the system – a matter for more discussion later.

The process of system modelling may highlight quite quickly measures which might be implemented immediately to reduce the occurrence or consequence of event or system failure. This is an important aspect of attempting to model a system since it forces the analyst(s) to understand in some considerable detail the system being considered. Such understanding may then make it clear what problems exist with the present system and this in turn might suggest useful amelioration measures.

More generally, however, system modelling is the first step in the quantitative analysis of system performance – which might be referenced to such measures as safety, mission success, availability and repair costs for the system. System modelling will help clarify which system elements or components need detailed quantification for system risk analysis. For example, quantitative information may be required on the reliability of pipes or valves, probabilistic models of earthquake loads, and estimation of operator error rates. This data, when incorporated into system representation logic diagrams, will enable the direct calculation of system risk.

Given that the system has been understood and modelled in an appropriate manner the next step is the computation of system risk. This requires knowledge of the performance of the components and subsystems. Conventionally such performance has been expressed through 'point estimate' values, single numerical values estimated from data or models of performance. However, as noted in Chapter 1, the data input to a system analysis will involve uncertainty in data. There will also be uncertainty in the models themselves. Rather than use point estimates, a better approach is to make specific allowance for uncertainty in data and in models. This suggests the use of random variables. Although there are many ways in which variability and uncertainty may be represented, the approach adopted herein is that used most widely in risk analyses, namely the use of probabilistic descriptions (probability density functions). The general format of such probabilistic measures is described later in the present chapter. The treatment of data and the development of probabilistic models will be described in Chapter 4 for specific system elements such as hardware items, resistances, strengths and loads and in Chapter 5 for human error. The stage is then set for quantitative system analysis in Chapter 6.

3.2 SYSTEM FAILURE

For most engineering risk analyses, system failure requires that there is a failure event and that this event causes the occurrence of undesirable consequences. In line with the usual definition discussed in Chapter 1, system risk may be defined as 'system risk' = 'failure frequency' × 'consequence of failure'. The failure event may be a release of latent energy (e.g. vapour cloud, explosion, fire, structural collapse, aircraft crash) or some other event (e.g. computer shutdown, poor-quality water), or more generally where the system does not perform according to its designed function. The (usually undesirable) consequences of a failure event may be categorized as:

1. safety hazard to personnel or to the public;
2. damage to the surrounding environment;
3. physical damage to the system; or
4. interruptions to the operation of the system (e.g. loss of production, downtime).

For example, system risk may be estimated for the probability or likelihood of being killed in an aircraft crash; for the probable loss of habitability of a geographic region after a radioactive release from a nuclear power plant; or for the probability of contracting an illness after drinking water which is potentially polluted.

Naturally, estimates of system risk obtained from a quantitative or probabilistic risk analysis will depend upon the nature and scope of the system and on the definition of what is considered to constitute 'system failure'. In general there will be a number of different failure modes which contribute to the 'system failure' criterion. What constitutes failure will also depend on the purpose(s) for which the risk analysis is carried out. Thus, estimates of system risk for hazards to people and the environment are used mainly to demonstrate compliance with regulatory safety goals or targets. On the other hand, estimates of risk for damage or interruptions to the system may help owners, managers and other decision-makers assess the efficiency or profitability of the system.

Not surprisingly, risk analyses often concentrate on risks (and consequences) that directly affect people; namely, loss of life, injury or economic losses. These consequences may be quantified using established statistical and probabilistic methods (e.g. Evans et al., 1985). However, translating loss of life or injury into monetary terms is a difficult and subjective task. The value of human life may be based, for example, on a range of measures such as future production, administrative decisions, consumer preference, court awards or life assurance. This topic is discussed in more detail in Chapter 7.

For some systems it is important that the consequences of system failure include damage to the environment. In this context, the 'environment' may be defined (DOE, 1991) as either the:

1. natural environment: national nature reserves; sites of special scientific interest; freshwater and estuarine habitats; aquifers and groundwater; marine environment; particular species; and the wider environment; or the
2. man-made environment: built heritage (buildings with architectural, historic or archaeological importance); and recreational facilities.

Damage to these environments may be immediate and may have direct effects. Immediate, delayed or permanent damage is also possible. In addition there may be indirect but harmful effects on people; these may include the contamination of crops, animals, and water supplies; and damage to hospitals and to sewerage systems. Guidelines on what constitutes damage to the environment are available. For example, a 'major accident' is defined for a national nature reserve as permanent or long-term damage in 'more than 10% or 0.5 hectares (whichever is lesser) of the area; more than 10% of the area of a particular habitat; or more than 10% of a particular species associated with the site' (DOE, 1991). This definition provide some basis for determining the nature of the environmental risk to be quantified. Codes of practice and guidelines to assist in this are also available (e.g. HSE, 1990; CIMAH, 1984).

3.3 IDENTIFICATION OF SOURCES OF RISK

The first step in a system risk analysis is the identification of the sources of risk to the system. Chapter 2 gave an overview of these for a range of systems. In being able to identify sources of risk it is essential for the analyst to be familiar with the system under consideration. Typically, a study or review team should be established. The team will comprise mainly managers, engineering staff, operators and other personnel who are involved in the operation of the system or who contribute to its performance. The range of knowledge and experience of the study team is a major factor in its effectiveness and hence in the competence of the study. However, it should be recognized that it may not be possible for the team to identify all possible failure scenarios or hazards particularly where these arise from 'unforeseen' events or processes (see section 2.3).

The following techniques have been used in the identification of sources of risk and will be described herein:

1. Preliminary Hazard Analysis (PHA);
2. Failure Modes and Effect Analysis (FMEA);
3. Failure Mode, Effect and Criticality Analysis (FMECA);
4. Hazard and Operability Studies (HAZOP); and
5. Incident Databanks.

These techniques have been used for a large range of engineering systems, with the possible exception of HAZOP which tends to be specific to the chemical and process industries. In addition, various other methods have been proposed, although these are often adaptions of the above methods to suit a specific system or problem. It will be seen that the methods to be described tend to be complementary; for example, guide lists, checklists or reference to incident databanks are often used to check that no source of risk has been omitted from the analysis.

3.3.1 PHA – Preliminary Hazard Analysis

A Preliminary Hazard Analysis (PHA) is used to identify the major hazards for a system, their causes and the severity of the consequences. Typically it is used at the preliminary design stage. However, it may also be used at later design stages to assess whether new hazards have been introduced (e.g. Henley and Kumamoto, 1981; Villemeur, 1991). The results of a PHA typically are presented in tabulated form (e.g. Table 3.1). Guide lists of elements that are potentially hazardous and lists of potentially hazardous situations for specific systems often aid the conduct of a PHA, as do checklists: see Table 3.2 for a typical checklist in the aeronautical industry. Another technique, used in the nuclear industry but being adopted elsewhere, is the so-called 'walk-down' procedure. This is an on-site, visual and methodical inspection and evaluation of the equipment in its installed condition under operational conditions, considering matters such as vibration, corrosion, containment etc. Evidently, it requires careful selection of appropriate criteria and the availablity of skilled and highly experienced personnel for the assessment process.

The identification in a PHA of major hazards usually will invoke more detailed analyses using methods such as FMEA, FMECA and HAZOP. Because of its preliminary status, it would not be expected that a PHA will identify failure of a specific individual component which has the

Table 3.1 PHA format, for a chemical process plant

Hazardous element	Event causing hazardous situation	Hazardous situation	Event leading to a potential accident	Potential accident	Effects	Preventative measures
1. Strong oxidizer	Alkali metal perchlorate is contaminated with lube oil	Potential to initiate strong redox reaction	Sufficient energy present to initiate reaction	Explosion	Personnel injury; damage to surrounding structures	Keep metal perchlorate at a suitable distance from all possible contaminants
2. Corrosion	Contents of steel tank contaminated with water vapour	Rust forms inside pressure tank	Operating pressure not reduced	Pressure tank rupture	Personnel injury; damage to surrounding structures	Use stainless steel pressure tank; locate tank at a suitable distance from equipment and personnel

Source: adapted from Henley and Kumamoto (1981).

Table 3.2 Checklists used in the aeronautical industry

Hazardous elements	Hazardous situations
Fuels	Acceleration
Propellants	Contamination
Initiators	Corrosion
Explosive charges	Chemical dissociation
Charged electrical capacitors	Electricity (shock, power failure)
Storage batteries	Explosion
Pressure containers	Fire
Spring-loaded devices	Heat and temperature
Suspension systems	Leakage
Electrical and gas generators	Moisture
Falling objects	Oxidation
Catapulted objects	Pressure
Heating devices	Radiation
Pumps	Mechanical shock
Blowers, fans	Vibrations, noise
Rotating machinery	
Actuating devices	
Nuclear devices	
Reactor	
Energy sources	

Source: Villemeur (1991). Reprinted by permission of John Wiley & Sons Ltd. Copyright ©1991.

potential to lead to a major hazard. This is the role for FMEA, FMECA and HAZOP.

3.3.2 FMEA – Failure Modes and Effect Analysis

The Failure Modes and Effect Analysis (FMEA) method was developed in the 1960s for aerospace applications (Recht, 1966), and is widely used in the aerospace, nuclear, electrical and electronics and manufacturing industries. In a FMEA, each component (or sub-task) in the system is systematically reviewed in order to identify:

1. the components or processes needed for successful operation of the system, their function, and operating states;
2. the failure modes of these components or processes – all possible ways a component can fail to perform its function (e.g. valve stuck open, stuck closed, partly open);
3. the causes of the failure modes – internal causes (e.g. equipment failure) and external causes (e.g. loss of electricity supply, operator error);
4. the method(s) of detecting and/or correcting these failure modes (e.g. inspections, maintenance); and
5. the effect of failure events on the performance of other components and on the system.

Table 3.3 Generic failure modes

1 Structural failure (rupture)	19 Fails to stop
2 Physical binding or jamming	20 Fails to start
3 Vibration	21 Fails to switch
4 Fails to remain (in position)	22 Premature operation
5 Fails to open	23 Delayed operation
6 Fails to close	24 Erroneous input (increased)
7 Fails open	25 Erroneous input (decreased)
8 Fails closed	26 Erroneous output (increased)
9 Internal leakage	27 Erroneous output (decreased)
10 External leakage	28 Loss of input
11 Fails out of tolerance (high)	29 Loss of output
12 Fails out of tolerance (low)	30 Shorted (electrical)
13 Inadvertent operation	31 Open (electrical)
14 Intermittent operation	32 Leakage (electrical)
15 Erratic operation	33 Other unique failure conditions as applicable to the system characteristics, requirements and operational constraints
16 Erroneous indication	
17 Restricted flow	
18 False actuation	

Source: Villemeur (1991). Reprinted by permission of John Wiley & Sons Ltd. Copyright ©1991.

FMEA is an inductive analysis because it starts at the possible outcomes and works backwards to obtain all possible causes. Hence it is essential that the identification of failure modes be as extensive as possible. This may be difficult, particularly for large systems. For this reason, generic guidelines or checklists are often used to ensure that all failure modes are considered, see, for example, Table 3.3. The analysis involved in a FMEA generally is presented in a tabulated format (see Table 3.4) in a manner similar to that used for a PHA.

Note that in a FMEA only one component is considered at a time, and it is assumed that all other components are functional (Aven, 1992). Thus, a FMEA normally will not reveal critical combinations of failure events or processes that will lead to system failure.

It has been claimed that FMEA is particularly time consuming because all failure events (including non-hazardous failures) must be considered (Leveson, 1986). However, the time effort associated with a FMEA may be reduced significantly if the analysis concentrates initially on sub-systems. If failure of a sub-system is deemed non-critical, then all components within that specific sub-system can be omitted from the detailed FMEA (Villemeur, 1991). Further, a systematic approach used in a FMEA may help in identifying hazards which might otherwise have been missed, overlooked or not imagined.

The FMEA approach, when used for identifying potential human errors in nuclear power plant operations, has been referred to as a 'Task Driven Method' (Cacciabue, 1988).

Table 3.4 Example of FMEA table

Component identification (identification code, name, type location)	Functions, states	Failure modes	Possible failure causes (internal, external causes)	Effects on the system	Detection means	Operator's action	Comments
Identification code: 031 VD Name: valve regulating feedwater flow to the Steam Generator (SG)1 produced by Motor-Driven Pump (MDP) 021P Type: regulating valve Location: KA 0524	Function: regulation of feedwater flow to SG1 produced by MDP 021P States: normally open valve	1. Valve stuck wide open	• Internal mechanical defect • Defect of control air system • Loss of control air to motor (SAR) • Loss of control power (125 V. Channel A)	• The flow rate supplied by SG1 by MDP 021P cannot be controlled from the control room • In case water or steam pipe breaks, or of SG tube break, SG1 cannot be isolated from control room	• Limit switch • SG1 supply flow rate • Exceptionally high flow-rate alarm (101 and 102 MD). • Possibly high flow-rate alarm (101 and 102 MD) threshold set 120 t/h	• The operator must position the valve locally • The operator must stop MDP 021P or close valve 051 VD locally	• The regulating valves operating with MDP are powered through the same channels as the pumps; the regulating valves operating with the turbine-driven pumps are powered through channels A and B

Source: Villemeur (1991). Reprinted by permission of John Wiley & Sons Ltd. Copyright ©1991.

3.3.3 FMECA – Failure Mode, Effect and Criticality Analysis

A Failure Mode, Effect and Criticality Analysis (FMECA), or simply Criticality Analysis, is a logical extension of FMEA in which failure events are categorized according to the seriousness of their possible effect. In the FMECA both the failure frequency (probability) and the failure effect (consequence) are assessed subjectively to determine the criticality of each failure mode. This should take account of each component and each subsystem. The failure frequency is rated in terms of a subjective likelihood such as expressed by 'very low, low, medium and high'. The severity is assessed into one of a number of subjective severity levels, for example (Leveson, 1986):

Level 1   minor (extra maintenance)
Level 2   significant (delay failure)
Level 3   critical (shutdown failure)
Level 4   catastrophic (potential loss of life)

The level of refinement desired in the analysis will determine the number of categories and their delineation. Critical failure modes are those with high failure rates and/or high levels of severity; these occur in the lower right-hand corner of the typical FMECA matrix presentation of results as given in Table 3.5.

An alternative and more quantitative approach is to estimate a 'criticality number' (C_m) for each component or sub-system. The criticality number represents the number of losses per reference time period (e.g. per year). It provides a means to rank components or sub-systems, and is calculated for each severity level (m) from:

$$C_m = \sum_{i=1}^{N} \beta_i \alpha_i \lambda_p t \qquad (3.1)$$

where N is the number of failure modes for the component or subsystem; β_i is the conditional probability that the loss (consequences or severity) of the failure mode will occur given that the failure mode has occurred (e.g. $\beta_i = 1.0$ represents actual loss); α_i is the failure mode ratio ($\Sigma \alpha_i = 1$); λ_p is the component failure rate in failures per hour or cycle;

Table 3.5 FMECA table

Severity	Probability			
	Very low	Low	Medium	High
Level 1 – minor				
Level 2 – significant				
Level 3 – critical				
Level 4 – catastrophic				

and t is the operating or 'at-risk' time of the component or sub-system in the reference time period (e.g. ARP-926, 1966; MIL-STD-1629).

Note that FMECA methods have no absolute meaning because they do not attach actual (or absolute) values to the consequences (e.g. lives lost, monetary damage). Instead, the FMECA approach uses a comparative (or sensitivity) approach with the failure modes ranked so as to identify those failure modes critical to the system. The failure modes which have been identified as critical components might then be subject to further study, or the components might be replaced by more reliable components, or the system or the critical components (or both) might be subject to increased monitoring (e.g. maintenance or replacement) or the system may need to be redesigned. Note that it is assumed in a FMECA that there is independence between the components or sub-systems, otherwise dependence will add significantly to the complexity of the analysis.

The FMEA and FMECA methods are described in a number of standards (e.g. ARP-926, 1966; MIL-STD-1629; IEEE 352, 1975; BS 5760, 1982) and extensively in the literature (e.g. Villemeur, 1991; Aven, 1992; O'Connor, 1991).

3.3.4 HAZOP – Hazard and Operability Studies

The HAZOP (Hazard and Operability Studies) technique was developed by Imperial Chemical Industries (Lawley, 1974) and is used widely in the chemical and in the process industries to identify hazards or operating problems in new or existing plants. HAZOP is an adaptation of the FMEA technique specifically for systems based on (pipeline) flows and process units.

The HAZOP technique is a systematic process in which process flow diagrams are used to consider each plant item (e.g. pipe, valve, computer software) in turn so that problems which could occur with these items may be considered. For example, issues raised may include the potential for reverse flow in a component (if so, is this hazardous?), the effect of changes in chemical concentration, and the influence of maintenance practices on items of equipment. Results from a HAZOP usually are summarized in tabular worksheet form. The tables normally contain the following entries:

1. Item: individual components in the system (e.g. pipes, vessels, relief valves)
2. Deviation: identify what can go wrong (e.g. more pressure, no transfer, less flow)
3. Causes: causes of each deviation (e.g. equipment failure, operator error)
4. Consequences: identify effects on other components, operability and

Identification of sources of risk

	hazards associated with each deviation (e.g. line fracture, backflow, leakage, fire, explosion, toxic release, personnel injury)
5. Actions:	measures or actions required to further reduce the deviations or the severity of the consequences (e.g. process design changes, equipment changes or modifications)

A typical HAZOP worksheet is given in Table 3.6. Guide words such as NO/NOT, MORE OF, LESS OF, OTHER THAN, AS WELL AS, PART OF and REVERSE typically are used to described deviations in flow, temperature, pressure, viscosity, concentration radioactivity and other process parameters. The use of guide words requires subjective estimates. Ideally, a HAZOP should be conducted as part of the plant design process. Further details of HAZOP are available; for example, Lawley (1974), Lees (1980), Kletz (1986) and Montague (1990).

3.3.5 Incident databanks

Accident data, 'near misses', reliability data and other statistics that describe the past performance of systems also may be used to help identify potential major hazards in a system, and their causes and their consequences. Further, such data may be used to supplement the combined knowledge and experience of study teams conducting PHA, FMEA, FMECA or HAZOP analyses. Care is required, however, since statistics may be easily misinterpreted (or even mis-used) and, since statistics are based on past events, may not be relevant to present or future situations.

Accident and 'near-miss' data are available in incident databanks. Component and system reliability databanks are also available, as discussed in Chapter 4. Some typical incident databanks include:

1. FACTS:	The Failure and Accidents Technical Information System collects data on world-wide industrial incidents (and near misses) for the production, storage, transport, use and disposal of chemicals (Bockholts, 1987).
2. SONATA:	Contains data on accidents during the storage, transport, extraction, handling and use of dangerous substances (Colombari, 1987).
3. AORS:	The Abnormal Occurrences Reporting System collects data on incidents in Nuclear Power Plants in Europe and the US (Kalfsbeek, 1987).
4. PLATFORM:	Contains data on accidents to drilling vessels and offshore oil platforms (Bertrand and Escoffier, 1989).
5. TANKER:	Contains data on accidents to ships that have caused an offshore oil spill of at least 500 tonnes (Bertrand and Escoffier, 1989).

Table 3.6 Example of HAZOP worksheet

Guide word	Deviation	Possible causes	Consequences	Action required
NONE	No flow	(1) No hydrocarbon available at intermediate storage	Loss of feed to reaction section and reduced output. Polymer formed in heat exchanger under no flow conditions	(a) Ensure good communications with intermediate storage operator (b) Install low level alarm on settling tank LIC
		(2) K1 pump fails (motor fault, loss of drive, impeller corroded away etc.)	As for (1)	Covered by (b)
		(3) Line blockage, isolation valve closed in error, or LCV valve fails shut	As for (1) J1 pump overheats	Covered by (b) (c) Install kickback on J1 pumps (d) Check design of J1 pump strainers
		(4) Line fracture	As for (1) Hydrocarbon discharged into area adjacent to public highway	Covered by (b) (e) Institute regular patrolling and inspection of transfer line
MORE OF	More flow	(5) LCV fails open or LCV bypass open in error	Settling tank overfills Incomplete separation of water phase in tank, leading to problems on reaction section	(f) Install high level alarm on LIC and check sizing of relief opposite liquid overflowing (g) Institute locking off procedure for LVC bypass when not in use (h) Extend J2 pump station line to 12" above tank base

Guide word	Deviation	Possible causes	Consequences	Action required
	More pressure	(6) Isolation valve closed in error or LCV closes, with J1 pump running	Transfer line subjected to full pump delivery or surge pressure	(j) Covered by (c) except when kickback blocked or isolated. Check line, FQ and flange ratings, and reduce stroking speed of LCV if necessary. Install a PG upstream of LCV and an independent PG on settling tank
		(7) Thermal expansion in an isolated valved section due to fire or strong sunlight	Line fracture or flange leak	(k) Install thermal expansion relief on valved section (relief discharge route to be decided later in study)
	More temperature	(8) High intermediate storage temperature	Higher pressure transfer line and settling tank	(l) Check whether there is adequate warning of high temperature at intermediate storage. If not, install
LESS OF	Less flow	(9) Leaking flange or valved stub not blanked and leaking	Material loss adjacent to public highway	(m) Covered by (e) and the checks in (j)
	Less temperature	(10) Winter conditions	Water sump and drain line freeze up	(m) Lag water sump down to drain valve, and steam trace drain valve and drain line downstream
PART OF	High water concentration in steam	(11) High water level in intermediate storage tank	Water sump fills more quickly Increased chance of water phase passing to reaction section	(n) Arrange for frequent draining off of water from intermediate storage tank. Install high interface level alarm on sump

Source: Lawley (1974). Reproduced with permission of the American Institute of Chemical Engineers. Copyright ©1974 AIChE. All rights reserved.

6. AEPIC: The Architecture and Engineering Performance Information Centre has collected data on the performance of engineered structures and buildings (Loss and Kennett, 1987).

3.4 SYSTEM REPRESENTATION

3.4.1 General approach

The techniques described above assist in the identification of those individual system elements (components and sub-systems) that are potentially hazardous. This information may be used in the development of a representation of the overall system in terms of logic diagrams. These identify the sequences or combinations of events or processes necessary for system failure to occur. As noted earlier, such system representation diagrams are an aid in the understanding of the behaviour of the system and hence may suggest, without formal risk analysis, obvious measures for reducing the risk of system failure. Of course, detailed understanding of the system and its representation in a logical fashion is required for quantitative analysis of the system. Such analysis also requires quantification of system element performance, a matter discussed in more detail in Chapter 4. The quantification of the overall system reliability is discussed in more detail in Chapter 6.

The essential (and most common) techniques used for schematic representation of a system (i.e. its 'modelling') are (1) fault trees and (2) event trees. A decision tree is a special case of event tree. Other methods, such as cause–consequence diagrams and reliability block diagrams, incorporate significant features of event tree and fault tree techniques (e.g. Lees, 1980) and will not be discussed. Fault trees and event trees have much in common. Whether one or the other or some combination is applied depends much on the preferences and practices within a given industry. Moreover, fault trees and event trees are complementary techniques; fault trees use deductive (backward-looking, top-down) logic and event trees use inductive (forward-looking, bottom-up) logic. Typically in applications, a combination of fault trees and event trees is used for the system representation. For example, the nuclear industry uses both the 'small event tree/large fault tree' and the 'large event tree/small fault tree' approaches; the delineation between these approaches is dependent upon the extent of event resolution and the preference of the analyst (IAEA, 1992). Both methods have the ability to provide insights into system behaviour which might otherwise not have been apparent.

For an offshore platform, for instance, a fault tree approach produced the following sequence of failure events (i.e. path through a fault tree):

1. rupture of first stage separator;
2. liquid carry-over to first stage compressor;
3. backflow from gas re-injection well;
4. liquid in fuel gas to turbine;
5. liquid carry-over to HP flare;
6. large unignited gas release via HP flare;
7. backflow from water injection well;
8. main electrical power failure;
9. failure of gas detection;
10. failure of ESD valve to close when demanded;
11. failure of firewater deluge system;
12. failure of Halon System to operate on demand;
13. failure of free-fall lifeboat launching system when demanded.

Each of these individual failure events may be caused by equipment failure or human error. The occurrence of all 13 failure events obviously would result in large losses of life and extensive damage to the deck. However, particular combinations of two or more failure events are less likely to occur but if they do occur could also lead to other forms of system failure; for example, the occurrence of failure events 1–9 would probably cause damage to equipment and/or deck, and loss of production (Slater and Cox, 1985).

Fault trees and event trees have been applied extensively for qualitative and quantitative risk studies in the nuclear industry and the chemical process industries and to a lesser extent elsewhere. Two landmark risk analysis applications of fault trees and event trees were the US nuclear safety study (RSS, 1975) and the UK Canvey study of chemical process industries (HSE, 1978). It is the main method recommended for US nuclear risk studies, in part because of its ability to model very complex accident sequences, including those involving dependency between events. Also, the work in the aerospace industry for human reliability analysis has made extensive use of event tree methods (Swain and Guttman, 1983).

Event tree and fault tree representation is most suited to systems (1) in which events represent discrete occurrences, (2) for situations in which the events are essentially continuously operating components (as in electrical circuits or hydraulic networks), or (3) when some or all of the components are dependent on the state of other components (such as in structural engineering systems or in systems with standby type components or others involving switching and sequential logic in operation).

Event trees and fault trees are best developed by a study team (i.e. a panel of specialists), and their discussions may be likened to a 'brainstorming' session where specific risks, events and scenarios and their control are suggested. It is at this stage that decisions can be made as to which risks are to be included and omitted from a risk analysis. In other words, the scope of the risk analysis can be defined.

There is no single method or algorithm for the construction of fault trees or event trees. Essentially, representation of the system being considered requires a step-by-step approach. Each such step will add to the construction of the fault tree or the event tree. Some care is required in fault or event tree construction, however. Thus, over-development of one branch of a tree without proceeding with the others, working down level by level systematically, can lead to system misrepresentation, to omission of important failure mechanisms, to an inappropriate balance between component failures and human errors, and to failure to recognize dependence of events (Aven, 1992). The following sections describe the functions of fault trees and event trees; their construction should then become readily apparent. The importance of dependency between system elements and system outcomes and its representation (and analysis) is also discussed.

3.4.2 Fault trees

Fault trees show the causal relationships between various events and use a logic which is essentially the complement of that used in event trees. A fault tree starts at a possible system failure mode (top event or undesirable event) and then works backwards to identify events (fault events) which may contribute to a failure event; these events are enumerated in logical sequences (event statements) through logical connections (logic gates). Such an analysis produces a tree-like structure having basic events at its extremities. The basic events are those for which failure data is available or which cannot be further dissected into more elementary events. The basic events are sometimes distinguished as 'initiating' or triggering events and 'enabling' events. An initiating event is the first failure event in an event sequence, while an enabling event (e.g. warning system disabled for maintenance) will cause a higher level failure when accompanied by a lower level failure (e.g. initiating event).

Thus, a fault tree is a Boolean logic diagram comprised principally of AND and OR logic gates. The output event of the AND gates occurs if the input events occur simultaneously and for an OR gate if any one of the input events occurs, see Figure 3.1. Other logic gates such as DELAY, MATRIX, QUANTIFICATION, COMPARISON and others may be used also (e.g. Villemeur, 1991). Event statements are represented by a number of symbols; for example, a rectangle represents the top event or an intermediate event, a circle represents a basic event, and a diamond denotes an undeveloped event that is not subdivided into a number of basic events because of lack of information or usefulness. An example of a fault tree for the structural safety of a railway bridge is shown in Figure 3.2. Another simple example of fault tree representation, in this case for power supply to an industrial works, is shown in Figure 3.3.

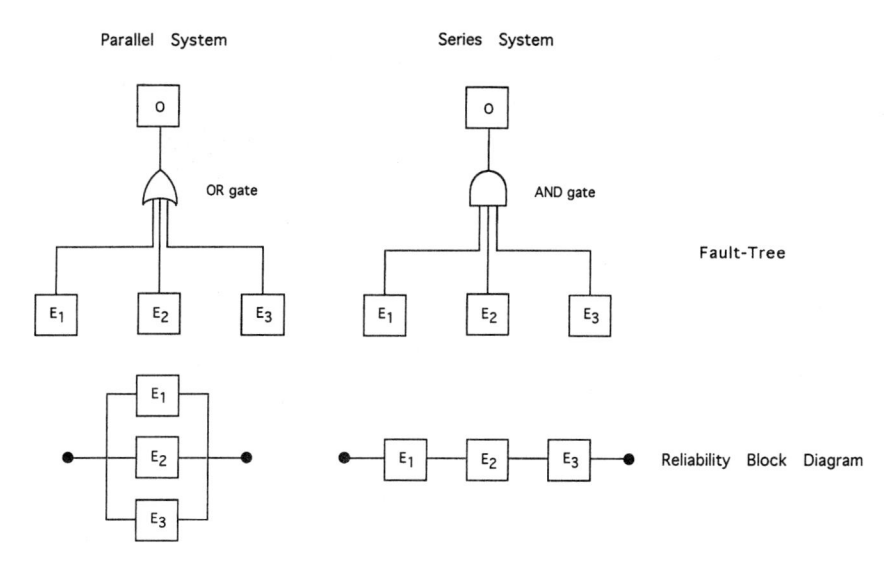

Figure 3.1 Principal Boolean logic for fault trees.

It is observed that a fault tree comprising an AND gate represents a 'parallel system'; namely, all components must fail for the system to fail. Hence, the system has a degree of redundancy because the system will still operate if one component fails. Conversely, an OR gate denotes a 'series system' in which all components must function for the successful operation of the system. In other words, a series system is analogous to a chain that is only as strong as its weakest element. Most engineering systems are a combination of series and parallel systems. A fault tree that comprises of AND and OR gates only can alternatively be represented by parallel and series reliability block diagrams, see Figure 3.1. Series and parallel systems are terms often used in system representation (e.g. Aggarwal, 1993; Melchers, 1987).

System failure modes are defined by a 'cut set' (or 'cut path'). A cut set is a combination of events (i.e. a section of a fault tree), in which the top event will definitely occur. A large number of cut sets may exist for a system; for example, hundreds of cut sets may exist for a system comprising between 40 and 50 components. Note that the top event in a cut set may still occur even though not all basic events in the cut set occur. A 'minimal cut set' is a cut set that represents the smallest event combinations that cause the top event to happen; it is essentially a 'critical path' (Villemeur, 1991). Thus, the top event will occur only if all events in the minimal cut set occur. A cut set has the same number or more events than a minimal cut set; this reduces the number of basic events in a cut set and so simplifies numerical analysis. Figure 3.4 shows a reliability block diagram representation of a system with five

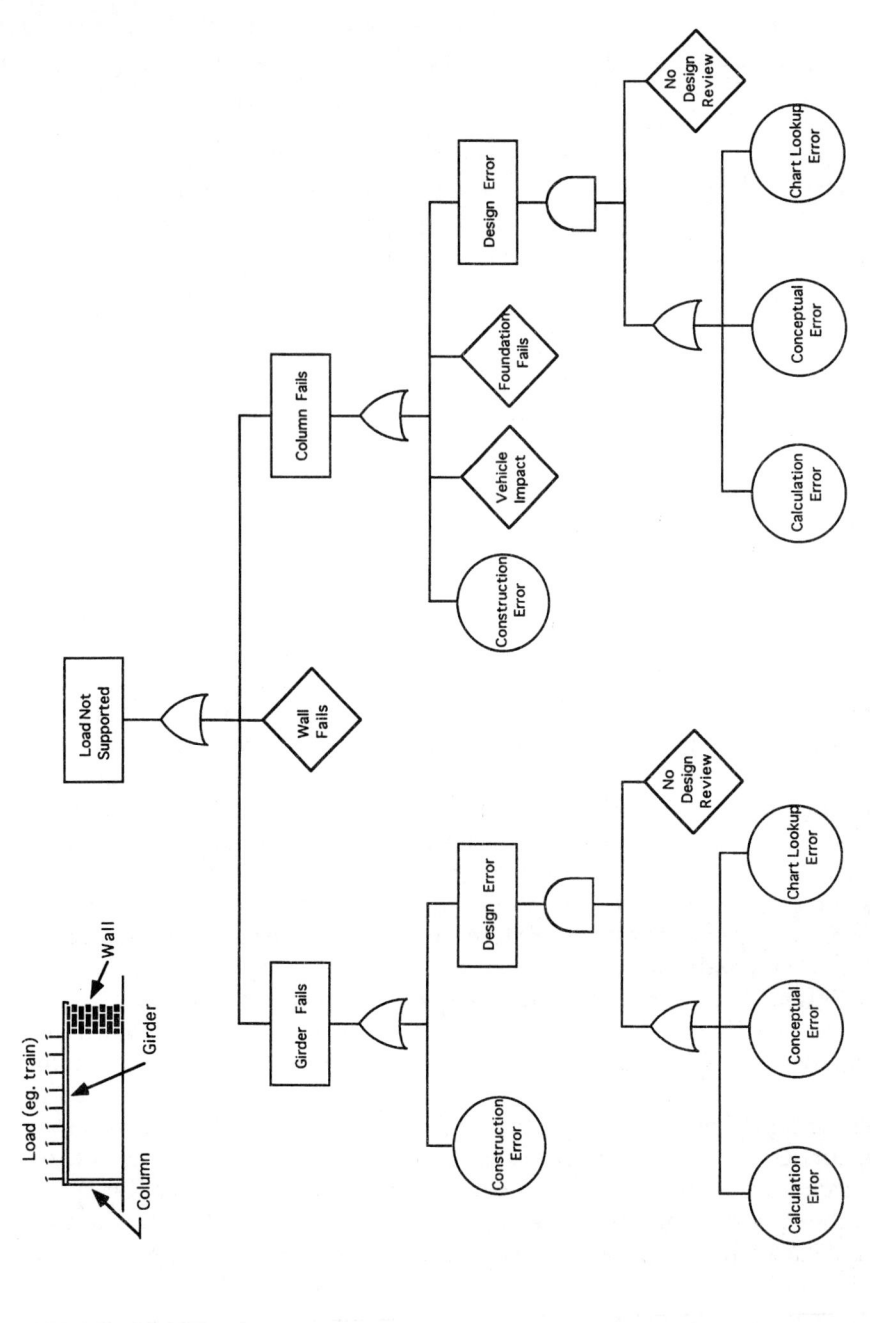

Figure 3.2 Fault tree for a railway bridge (assumes no train overload).

System representation

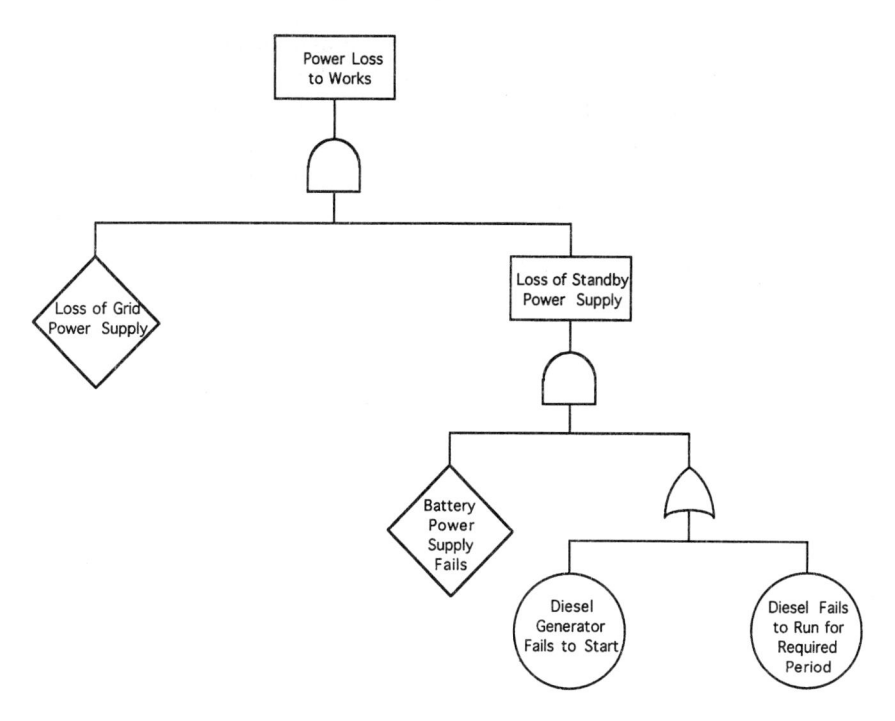

Figure 3.3 Fault tree for loss of works power supply.

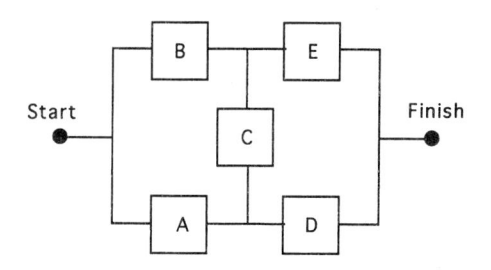

Figure 3.4 Reliability block diagram for a system comprising five components.

independent components; the minimal cut sets are AB, DE, ACE and BCD.

An alternative way of thinking about the problem is in terms of 'path' sets. Whereas cut sets identify failure modes of a system, minimal path sets identify survival modes. Thus in Figure 3.4 the minimal path sets are AD, BE, ACE and BCD. Hence the minimal path sets form the complement to minimal cut sets.

The 'top' event in a fault tree can occur only as a result of the occurrence of one or more basic events. Typically these are the components of the system and would include items such as hardware, various subsystems, environmental factors as well as personnel (human error) and social matters. Furthermore, various external events such as natural hazards, or man-made hazards may need to be considered. The components may fail during the operation, testing or maintenance of the system. Hence it is necessary to examine their operation or condition. Component failures are sometimes classified as 'primary' when the component itself fails and needs replacement, 'secondary' when the component fails due to excessive demands made upon it (i.e. beyond that for which it was designed) and 'command' when the component is not subject to the correct instruction, command or environment to allow it to perform in its intended manner. Sub-systems may be subjected to a separate fault tree analysis; this may occur at a later date when more refined knowledge of the system is available.

If adequate knowledge about the system exists and it can be formulated, it may be possible to use knowledge-based expert systems to help identify top events and the main initiating events (Hauptmanns, 1988). In addition, computer-based routines exist to help develop fault trees given connectivity of the process or system under consideration and the input-output relations for each event (e.g. Lee et al., 1985). Some of these routines generate minimum cut sets at the same time.

For some systems, one or more basic events may be mutually exclusive, that is, it may have at least two states of which only one can occur at any time (e.g. pump normal, pump stops). Such events complicate the construction of fault trees and the determination of cut sets and path sets. Further, the Boolean logic associated with fault trees considers only two outcomes from an event (e.g. failure and success). Another disadvantage of fault tree logic is that the possibility of a hazard and its consequences must be foreseen before a fault tree can be constructed (Lees, 1980), thus a preliminary study (or studies), using FMEA, FMECA, HAZOP or some other method, needs to be conducted prior to the development of the fault tree.

Further features of the fault tree method are described by Henley and Kumamoto (1981); Hauptmanns (1988); Lee et al. (1985) and Hadipriono and Toh (1989). Fault trees are also known as 'cause trees'.

3.4.3 Event trees

An event tree represents pictorially the logical order in which events in a system can occur. It commences with an initiating (or basic) event and works forward in time considering all possible subsequent events until the consequences are known – either the system corrects itself or some level of system failure occurs. Note that a number of possible consequences may arise from an event tree. These may include various levels or degrees of damage, personal injury or economic loss. An event tree

System representation

is constructed from event definitions and logic vertices (i.e. outcomes of an event); outcomes of events may produce more than two logic vertices. Initiating or basic events may be obtained from fault trees, FMEA, FMECA, HAZOP or other methods for identification of sources of risk. The general characteristics of an event tree are shown in Figure 3.5. A simple event tree for power supply to an industrial works is shown in Figure 3.6.

As noted in section 3.4 often a combination of fault trees and event trees is used for the representation of a system. Thus, the top event in a fault tree might become the basic event in an event tree or, alternatively, the consequence of an event tree might become the basic event in a fault tree (either for the system as a whole or for a sub-system). The first case might occur, for instance, if a pre-accident sequence is modelled as a fault tree and the post-accident sequence is modelled as an event tree. Figure 3.7 provides an example of this, where the consequences of an accident (bridge collapse) are represented as an event tree and a fault tree is used to describe the events leading to bridge collapse, as was earlier considered in Figure 3.2.

Event trees can become very complex. For example, for the system shown in Figure 3.4 the number of event tree paths is 32 (= 2^5) for the complete event tree, as shown in Figure 3.8. More generally, it is easy to show that for a system with n two-state components the total number of paths is 2^n and that if each component can have m states the total number of paths is given by m^n. The enumeration of all of these paths can be a considerable task. Fortunately, an event-tree can be simplified considerably and for many purposes such a 'truncated' or 'reduced' event tree is sufficient. Simplification is achieved by considering the possible outcome of the event sequence being worked on as each new component is added to the event tree sequence. If it is evident that only one outcome can be achieved then there is little point in continuing to develop the particular event path any further. For example, if components A and B have failed in the system shown in Figure 3.4 then it is clear that system failure has occurred. It follows that the event tree can be simplified as shown in Figure 3.9. Fortunately, in many cases only system failure is of interest for calculation purposes and a 'truncated' or 'reduced' event tree is sufficient. In such cases all paths leading to system success can be terminated as soon as it evident that this will be the outcome for the path. Where system evaluation proceeds in parallel with event tree construction, it may be possible to terminate the event sequence on the basis of the estimated probability of occurrence of the sequence (see Chapter 6).

As was the case for fault trees, cut sets and minimal cut sets are also applicable to event trees. For large and complex systems the minimal cut sets cannot be identified readily by inspection. Several methods for minimal cut set identification of networks have been developed for event

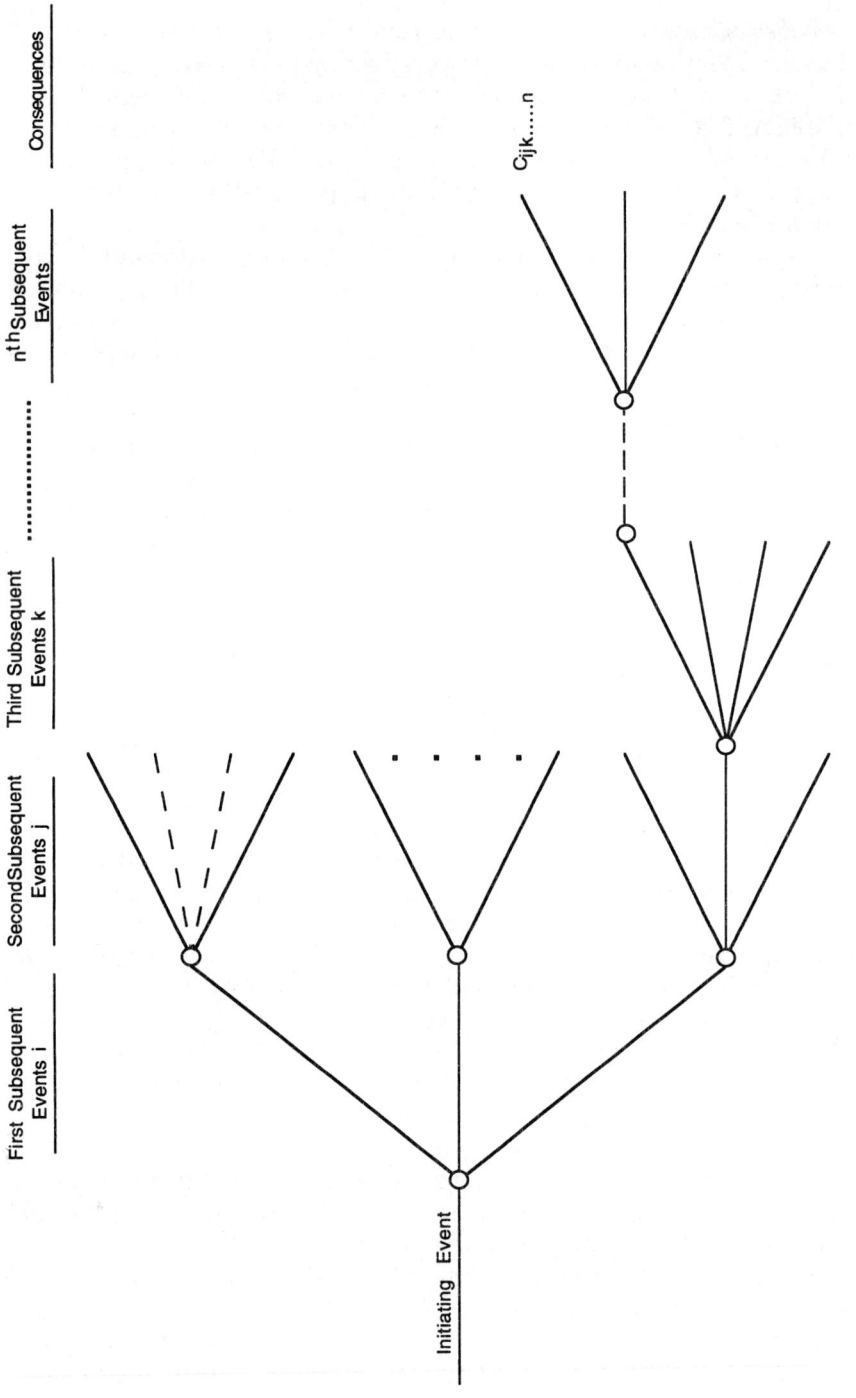

Figure 3.5 Event tree model

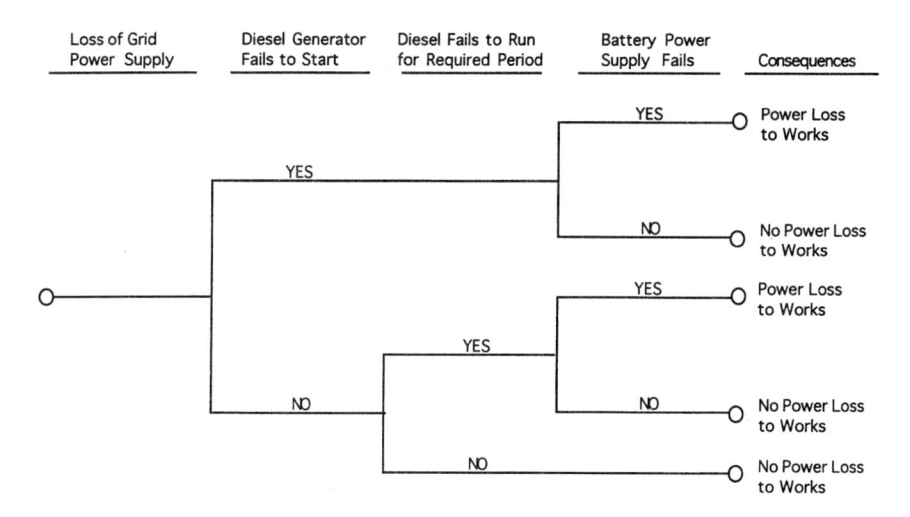

Figure 3.6 Event tree for loss of works power supply
Source: modified from Lees (1980).

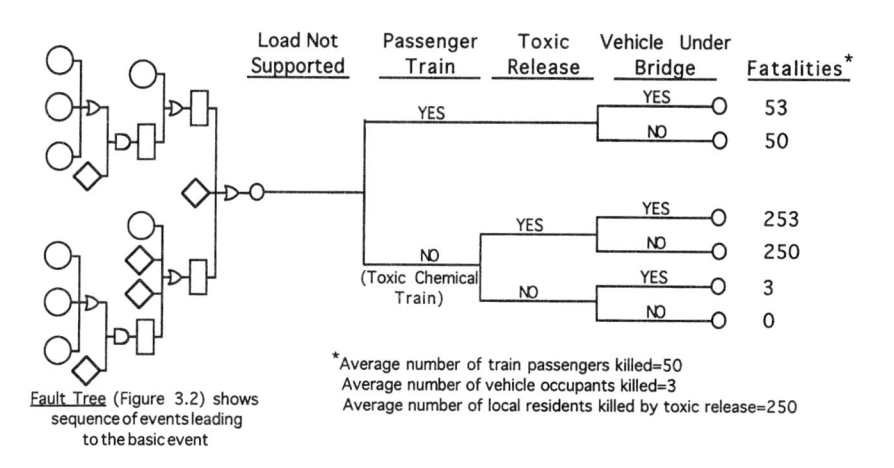

Figure 3.7 Event tree for consequences of railway bridge collpase.

*Average number of train passengers killed=50
Average number of vehicle occupants killed=3
Average number of local residents killed by toxic release=250

trees (e.g. Allan *et al.*, 1981) but are not necessarily very efficient for fault trees. Further details on event trees may be obtained from Ang and Tang (1984) and Villemeur (1991), the latter also listing a number of commercially available computer codes. Finally, note that a decision tree is a special category of event tree and that event trees are also known as 'consequence trees' and 'accident sequence trees'.

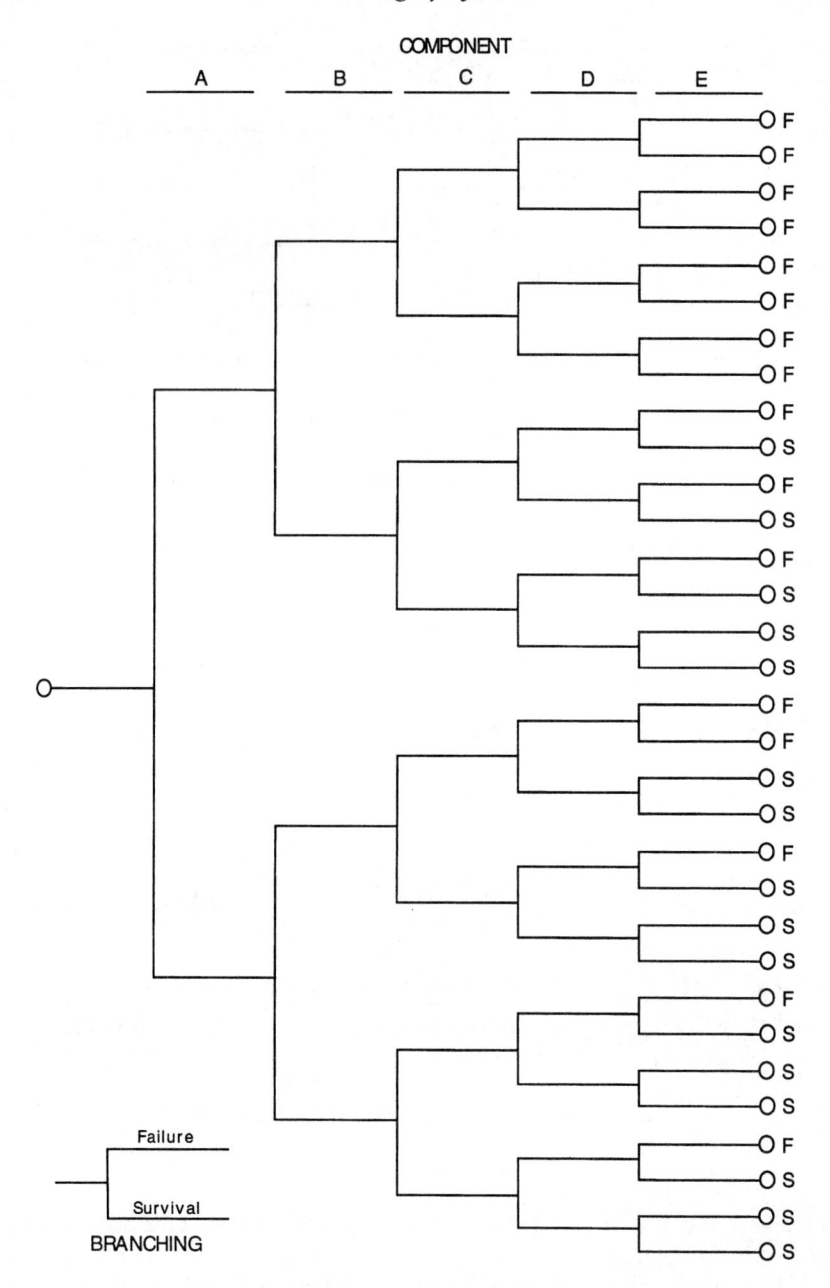

Figure 3.8 Event tree for system shown in Figure 3.4 (S = success, F = failure).

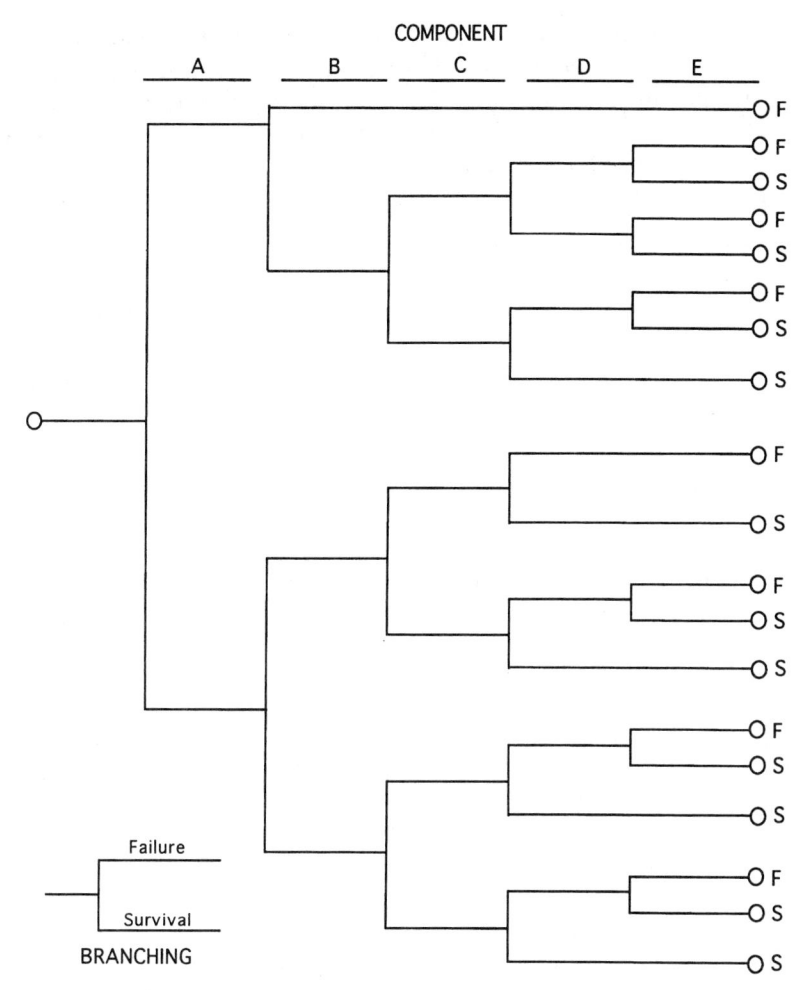

Figure 3.9 Truncated event tree (S = success, F = failure).

3.4.4 System dependency – common cause failures

It is sometimes assumed, for ease of analysis, that any dependence between the outcomes of events may be ignored. This assumption usually is incorrect. System risk estimates calculated on the basis of assumed complete dependence between events may be considerably greater than estimates determined for the same events being assumed to be completely independent. Dependence between event failures (also known as cascade failures) can occur when more than one component in a system fails simultaneously due to a common cause. In this case the components do not fail independently of each other. For example, an external agent (such as environmental load – wind, earthquake – or man-imposed factors) may affect more than one component in the system. In general, dependent failures can have a dramatic effect on the risk associated with a project and must be properly identified and accounted for. It follows that the treatment of dependency between events is an important matter for risk analysts.

Dependency modelling is often difficult because of:

1. inconsistent definitions of dependent events;
2. the lack of useable dependent event data;
3. a lack of appreciation of the importance and frequency of these events; and
4. the mistaken belief that the dependence problem is solved simply by doing a good job of explicitly modelling dependent events down to the component level.

In addition, modelling the causes of component failure and treating dependency between inputs which affect the response of a component may also be problematic (Fleming et al., 1986).

Dependency is particularly likely to occur for similar, multiple components. This is a result of (1) physical dependence, such as where the physical failure process(es) are related in some way (e.g. where there is dependence of the failure times of components involved in a multiple failure), and (2) statistical dependence, where the statistical properties are somehow correlated (e.g. Virolainen, 1984). It is also possible to have so-called 'state of knowledge' dependency due to a (common) lack of knowledge affecting more than one component or sub-system. Typically it is exhibited by similar statistical evidence of the failure behaviour of or in a system (Apostolikas and Kaplan, 1981). A somewhat similar dependency can arise through modelling assumptions or simplifications.

In the terminology adopted mainly in the nuclear industry, dependent failures are termed 'common cause failures' or, less frequently, 'common mode failures'. In the discussion to follow no distinction between these two terms will be made, although it is sometimes considered that the latter is a subset of the former.

A common cause failure situation is where an event is instrumental in triggering more than one following event without the analysis making specific allowance for this possibility. As noted, typical common cause events might be external agents such as natural or environmental causes or where failure of a piece of equipment or a supply system (e.g. electrical power failure) leads to failure of a number of dependent pieces of equipment. An example is the failure of both the main and the standby pumping systems due to electric power failure. This means that the failures are mutually dependent in some manner. A classification of sources of common cause failures is shown in Table 3.7.

Causes of common cause failures may be identified also from minimal cut sets. Each minimal cut set is assessed in order to determine if the

Table 3.7 Classification of common cause failures

Design		Construction		Procedural		Environmental	
Functional deficiencies	Design realisation	Manufacture	Installation and commissioning	Maintenance and test	Operation	Normal extremes	Energetic events
Logical error	Channel dependency	Inadequate quality control	Construction errors	Imperfect repair	Operator errors	Temperature	Fire
Inadequate measurement	Common operation and protection components	Inadequate standards	Inadequate quality control	Imperfect testing	Inadequate procedures	Pressure	Flood
						Humidity	Wind
Inadequate control		Inadequate inspection	Inadequate standards	Imperfect calibration	Inadequate supervision	Vibration	Earthquake
Inadequate response	Operational deficiencies	Inadequate testing	Inadequate inspection	Imperfect procedures	Communication error	Acceleration	Explosion
	Inadequate components		Inadequate testing and commissioning		Inadequate supervision	Stress	Missiles
	Design errors					Corrosion	Electrical power
						Contamination	Radiation
	Design limitations					Interference	Chemical sources
						Radiation	

Source: adapted from Edwards and Watson (1979).

performance of any of the events is subject to a common cause. However, this approach is only feasible for relatively simple systems; otherwise, the number of minimal cut sets to be assessed becomes prohibitively large.

Although, in principle, dependency in systems can be handled by appropriate detailed modelling of the system, this is not always possible. Very detailed system representation is not always possible due to insufficient understanding of the dependencies involved, or such detailed modelling not being desired due to the large amount of work involved. Thus simplified treatments of dependency tend to be used for many analyses but they are appropriate only where the effect of simplification on system outcomes is clearly recognized. This is not usually the situation in practice, where lack of data and other factors often dictate the use of simplified methods. Unfortunately, the short-cut methods for system dependence in the risk calculations usually are far from accurate.

In principle, fault trees can include the effects of common cause failures. In practice, however, the existence of common cause failure, and other sequential or time dependent events and, perhaps, the existence of control (or feedback) loops renders the construction of a fault tree very difficult. The reason for this is that the order of events is important (e.g. one component must fail before another can fail). In this situation the internal state of the system determines the occurrence of system failure and this is now a function of the state of the control loop(s). The problem is less severe for steady-state situations.

Event trees on the other hand, are probably more helpful in dealing with dependence. They trace the effect of the events, such as component failure, and this allows the propagation of the effects of common cause failure to be followed. Thus, the event tree technique is particularly useful where there is dependence between components such as that associated with common cause and other sequential type dependencies. It is evident that in this case the sequence or timing of events in the system must be carefully considered and properly reflected in event tree construction. This is not required in simpler systems with independent components, since then the sequencing of events can be more arbitrary.

One approach for handling common cause failures which has been adapted in the nuclear industry consists of assuming independence of all system events and introducing new, fictitious, events to deal specifically with the effect of dependency. For example, if an external event will lead to the simultaneous failure of parallel components, then the external event may be modelled as a fictitious event in series with the components, see Figure 3.10. Some further discussion of this topic is given in Chapter 6.

Finally, it should be evident that defining the impact of dependent events on the logic model should be one of the most important areas of

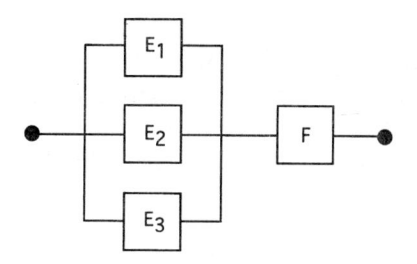

Figure 3.10 Parallel system with common cause failure event.

attention for analysts. Yet it would seem that there is often a considerably lower level of resources allocated to the analysis of dependent events compared to the analysis of the system as a whole and to the data analysis for independent events. This may lead to important system failure modes being overlooked and the system failure probability being not well estimated.

3.5 SYSTEM ELEMENT PERFORMANCE

3.5.1 Quantitative description

The estimation of system risk requires the quantitative description of both the frequency and the performance of those system elements directly influencing system risk. This means that the performance of components, items of equipment, loads, resistances, and human actions must be known and the consequences of failure able to be estimated. The quantitative description of the performance of each system element usually will be as a variable; either a point estimate (i.e. deterministic) variable (e.g. mean failure rate) or a random variable (e.g. probability distribution of failure rates). Broadly, variables may be categorized as:

1. reliabilities (failure rates, mean time to failure) of sub-systems, components or other items of equipment (e.g. pumps, valves, computers);
2. resistance, strength or capacity of mechanical, structural, geotechnical, electrical/electronic or hydraulic systems;
3. loads, stresses or demands placed on these systems (e.g. mechanical stress, voltage, internal stresses due to temperature changes, or structural loads due to wind and dead loads etc.);
4. human reliabilities (error rates, error detection rates) for administration, design, assembly/construction, operation, inspection and maintenance tasks; and
5. consequences of failure events such as the influence on other event performances, loss of life, economic losses etc.

As already noted earlier, the models used to describe component reliabilities, resistances, loads and consequences in a quantative manner are discussed in Chapter 4. The quantification of human reliabilities is considered in Chapter 5. In both cases it is important to recognize that the variables may be subject to several sources of uncertainty; specifically, these include:

1. inherent variability of the phenomenon itself (e.g. wind forces);
2. parameter;
3. prediction; and
4. modelling errors.

These matters, together with point estimate and random variables, and their ability to incorporate sources of variability and uncertainty, are described briefly in the following sections.

3.5.2 Variables as point estimates

A 'point estimate' is a single numerical value used to describe the best estimate of the value of a variable. For example, when the failure rate of a valve is estimated to be 2.4×10^{-6} failures/yr this value is a 'point estimate'. Typically, it is then assumed for system analysis that the reliability of all similar or identical valves is equal to 2.4×10^{-6} failures/yr at any time during the life of the system. However, in practice it is highly likely that component reliabilities will be influenced by the operating environment (humidity, pressure, dust), operating stress and maintenance practices. It is more likely, therefore, that the failure rate even for nominally identical components will vary from component to component. Further, the data used to calculate failure rates (e.g. laboratory testing, field data, expert opinions) is itself subject to some degree of uncertainty. All such variability is ignored in a point estimate.

A further example of a point estimate is the number N of people killed by an offsite toxic release. It is likely that N will be influenced by wind speed and direction, date and time of day, and effectiveness of evacuation procedures and other factors. Since point estimates for variables ignore variability and uncertainty it is not possible, easily, to obtain estimates for the uncertainty in the estimate of system risk (except to some extend through the use of sensitivity analysis – see Chapter 6). For this reason, it is often more realistic and more helpful for the failure rates for components and for the other variables such as resistances, loads, consequences, etc. to be represented as random variables.

3.5.3 Random variables

Variability and uncertainty in the value(s) associated with a variable arise from a variety of causes, including:

1. variations inherent in nature;
2. lack of understanding of all causes and effects in systems;
3. lack of sufficient data; and
4. workmanship.

For example, the strength of nominally identical components generally will vary from component to component due to variabilities in material properties, in production processes, in geometric dimensions, in environmental conditions, in maintenance etc. Although material properties may be considered time-invariant for many purposes, allowance may need to be made for wear, fatigue and corrosion. Loadings or stresses are influenced by a variety of factors, such as environment, temperature and geographic location. Many loading variables also tend not to be time invariant; for example, wind velocities are time-dependent. In this case, the load at any 'point-in-time' is likely to be highly variable from one time point to the next. Since in many cases the value of a variable cannot be predicted with a high degree of certainty, it is appropriate to consider the variable as a random variable and to treat it in a probabilistic manner. This means that the random variable will be described by a probabilistic model; typically this comprises a probability distribution, the form of which is described by one or more statistical parameters such as the mean and variance.

In developing and using probabilistic models, allowance must, obviously, be made for the uncertainties associated with each random variable in the model. The uncertainties which need to be considered can be categorized as:

1. natural variability of the probabilistic phenomenon itself;
2. modelling uncertainty; and
3. statistical parameter uncertainty.

Each of these uncertainties will now be described.

The inherent variability of a phenomenon is that which can be termed its natural (or intrinsic, irreducible, fundamental) uncertainty. For example, there is some natural uncertainty associated with the estimation of failure rates because it is generally accepted that there is a variation in failure rates even between identical components. In most cases, a probability distribution is used to represent this form of uncertainty. It therefore follows that the collection and analysis of more data will (in most cases) only slightly decrease the magnitude of natural uncertainties. The natural uncertainty aspect causes some additional problems when the prediction of future events (e.g. cyclones, earthquakes, floods, equipment failure) is of interest. This is because future events are not always directly related to historical data (i.e. past experience). This means that natural uncertainties may be particularly important when attempts are made to predict the occurrence of events beyond this data range. For example, it is likely that it will never be possible to predict accurately the precise timing or magnitude of maximum floods, even if the probabilistic models and their parameters are precisely known (i.e. perfect fit with experimental data).

It is reasonable to expect that the uncertainties associated with the occurrence and magnitude of natural phenomena (e.g. floods, cyclones, soil properties) are generally greater than those associated with man-made materials or processes (e.g. structural steel, equipment reliability).

Modelling uncertainty is concerned with the ability of probabilistic models to describe the relevant characteristics of system elements. Vesely and Rasmuson (1984) have categorized modelling uncertainty as (1) uncertainty as to whether all factors that influence the model are included, and (2) uncertainty in how the model describes the relationship between these factors. For example, it is often assumed that the occurrence of events may be modelled as a stochastic (i.e. random) process; however, the influence of long-term or other effects (e.g. 'El Nino' – or Southern Oscillation – phenomenon) may also need to be considered. Uncertainties of this type may become evident when the tails of proposed probabilistic models do not fit closely the rare events obtained in experimental data. In some cases, refinement of the model may be possible, particularly if the data are scarce and understanding is poor. Thus it is likely that modelling uncertainty will decrease as more data is collected and analysed.

The statistical parameters used to describe probabilistic models (e.g. mean, variance) are subject to some uncertainty themselves as they are estimated from limited amounts of experimental data. In particular there may be practical difficulties in collecting data for rare failure events or in eliciting high-quality expert opinion. Further, statistical parameters are also often obtained from generic data sources or from an extrapolation of data collected from other geographic locations, systems, or operating environments. Clearly, the values assigned to statistical parameters used in a probabilistic model are only estimates of the true values of the parameters. The collection and analysis of additional (and perhaps more appropriate) data ususally will lead to a reduction in statistical parameter uncertainty.

In describing random variables it is useful to note, for the discussion to follow, that random variables (and hence their probabilistic models) may be discrete or continuous. A discrete random variable is one in which there is possible only a limited (discrete) number of outcomes. One obvious example is the number of days. A continuous random variable is one for which the variable may take on any value within a range of possible outcomes. A simple example is the value of the temperature, such as 56.45 °C.

As will be known from statistics, the statistical parameters used to describe probabilistic models are the so-called first, second and higher 'moments'. The first and second moments are the well known mean (μ_X) and the variance (σ_X^2) respectively. The mean and variance for discrete random variables (x) are:

$$\mu_X = \sum_{\text{all } x_i} x_i p_X(x_i) \qquad \sigma_X^2 = \sum_{\text{all } x_i} (x_i - \mu_X)^2 p_X(x_i) \qquad (3.2)$$

and for continuous random variable:

$$\mu_X = \int_{-\infty}^{\infty} x f_X(x) dx \qquad \sigma_X^2 = \int_{-\infty}^{\infty} (x - \mu_X)^2 f_X(x) dx \qquad (3.3)$$

where $p_X(x_i)$ is the likelihood of occurrence of each discrete value x_i and $f_X(x)$ is the probability density function for the continuous variable x (e.g. Benjamin and Cornell, 1970; Ang and Tang, 1984). Evidently, the mean provides the best estimate of the value of the random variable, that is, it is the value most likely to occur in practice. The variance provides a measure of the uncertainty associated with the random variable. Other measures of this are the standard deviation (s_X) and the coefficient of variation ($V_X = \sigma_X/\mu_X$). The third moment is the 'skewness' of the distribution and provides a useful measure of its asymmetry. A large number of discrete and continuous probability distributions have been developed, including: binomial, exponential, poisson, normal, lognormal, gamma, exponential, extreme value (Type I, II or III), beta and others. In each case the 'moments' are the parameters of the distribution – although more parameters are required to describe some distributions than others. Figure 3.11 shows some typical continuous probability distributions, each with identical statistical parameters. The choice of probability distribution is dependent upon the characteristics of the random variable and the closeness with which the analyst wishes to describe it for risk analysis purposes. For example, lognormal distributions are used to model error rates because it has been observed that the performance of skilled persons tends to bunch up towards low error rates (Swain and Guttman, 1983). Also, the annual maximum wind speeds have been described by an extreme value type I (or Gumbel) distribution although other distributions also offer reasonable fits to the data (e.g. Simiu et al., 1978). Figure 3.11 shows how the selection of probability distribution can influence the characteristics of a random variable.

In most cases, uncertainties are incorporated directly into probabilistic models by simply increasing the magnitude of its variance. Uncertainties also may be incorporated into probabilistic models by treating the statistical parameters of the probabilistic model (e.g. mean, variance) as random variables; the probabilistic model then becomes a compound distribution (e.g. Benjamin and Cornell, 1970).

Finally, in many cases it is not possible to include all sources of uncertainty in the probability models. These sources of uncertainty are

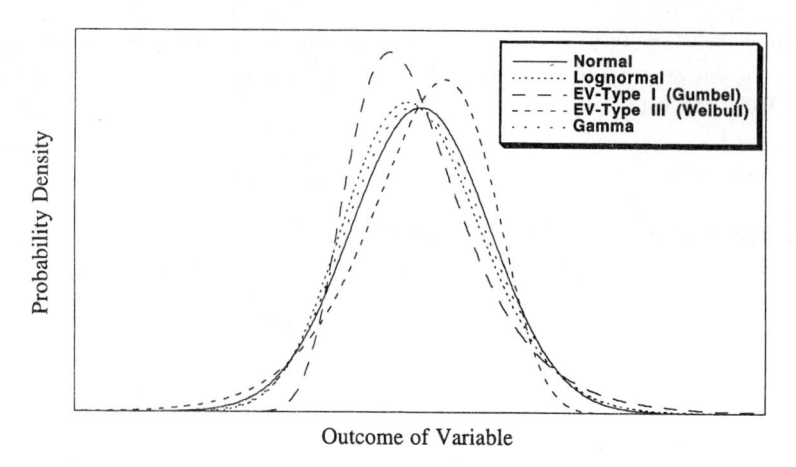

Figure 3.11 Typical continuous probability distributions.

essentially non-quantifiable and are usually associated with matters such as possible bias of system analysts, the expertise of the study team, the possible exclusion of some failure events etc. Quality assurance measures and peer review should address these issues and enhance the credibility and accuracy of the analysis, including amelioration of these and other sources of uncertainty. This aspect is discussed in more detail in Chapter 7.

3.6 UNCERTAINTY IN ANALYSIS OUTCOMES

It is clear from the above that there is variability and uncertainty associated with, among other things, the prediction of future events, time-dependent changes, experimental data and human error. The use of random variables provides an appropriate, probabilistic description of uncertainty. Since a risk analysis should also be concerned with uncertainty and variability in the results predicted by the analysis, it is appropriate for a risk analysis to employ random variables in the analysis. When this is done, the outcome of the analysis (e.g. system risk) is actually a dependent variable, which may be represented by a probability distribution. The spread of this distribution provides a direct measure of the uncertainty in the outcome of the analysis. As already noted, the use of point estimates variables in the analysis will lead to a point estimate outcome. This provides no indication of variability or uncertainty in the outcome.

3.7 SUMMARY

System representation using event tree or fault tree logic diagrams defines the scope of the system (and study) and identifies system elements that contribute to system failure. Common cause and other dependent failures are important contributors to system risk so care must be taken when incorporating these failure events into system representation. This information, when combined with system element performance data, is used in system evaluation models to calculate system risks. Naturally, system failure needs to be defined and sources of risk identified if system representation is to be accurate. It is often more meaningful for system element performance to be modelled as random variables (i.e. probabilistically) since system element performance is subject to variability and its prediction is uncertain also.

REFERENCES

Aggarwal, K.K. (1993) *Reliability Engineering*, Kluwer Academic Publishers, Dordrecht, The Netherlands.

Allan, R.N., Rondiris, I.L. and Fryer, D.M. (1981) An efficient computational technique for evaluating the cut/tie sets and common-cause failures of complex systems, *IEEE Transactions on Reliability*, **R-30**(2), 101–9

Ang, A. H-S. and Tang, W.H. (1984) *Probability Concepts in Engineering Planning and Design: Volume II – Decisions, Risks, and Reliability*, Wiley & Sons, New York.

Apostolikas, G. and Kaplan, S. (1981) Pitfalls in risk calculation, *Reliability Engineering*, **2**, 135–45.

ARP-926 (1966) *Design Analysis Procedure for Failure Mode, Effects and Criticality Analysis*, Recommended Practice ARP-926, SAE Aerospace.

Aven, T. (1992) *Reliability and Risk Analysis*, Elsevier Applied Science, London.

Benjamin, J.R. and Cornell, C.A. (1970) *Probability, Statistics, and Decision for Civil Engineers*, McGraw-Hill, New York.

Bertrand, A. and Escoffier, L. (1989) IFP databanks on offshore accidents. In V. Colombari (ed.), *Reliability Data Collection and Use in Risk and Availability Assessment*, Springer-Verlag, Berlin, 115–28.

BS 5760 (1982) *Reliability of Systems, Equipment and Components*, British Standards Institute, London.

Bockholts, P. (1987) Collection and applications of incident data. In A. Amendola and A.Z. Keller (eds), *Reliability Data Bases*, D. Reidel Publishing Company, Holland, 217–31.

Cacciabue, P.C. (1988) Evaluation of human factors and man-machine problems in the safety of nuclear power plants, *Nuclear Engineering and Design*, **109**, 417–31.

CIMAH (1984) *The Control of Industrial Accident Hazard Regulations (CIMAH)*, Statutory Instrument 1984/1902.

Colombari, V. (1987) The ENI databanks and their uses. In A. Amendola and A.Z. Keller (eds), *Reliability Data Bases: Proceedings of the ISPRA Course held at the Joint Research Centre, Ispra, Italy*, D. Reidel Publishing Co., Holland, 243–58.

DOE (1991) *Interpretation of Major Accident to the Environment for the Purposes of the CIMAH Regulations,* A Guidance Note by the Department of the Environment, Toxic Substances Division, Department of the Environment, London.

Edwards, G.T. and Watson, D.W. (1979) *A Study of Common-Mode Failures,* United Kingdom Atomic Energy Authority, SRD R 146.

Evans, J.S. et al. (1985) *Health Effects Model for Nuclear Power Plant Consequence Analysis,* Harvard University, NUREG/CR-4214, SAND85-7185.

Fleming, K.N., Mosleh, A. and Deremer, R.K. (1986) A systematic procedure for the incorporation of common cause events into risk reliability models, *Nuclear Engineering and Design,* **93**, 245–73.

Hadipriono, F.C. and H.S. Toh (1989) Modified fault tree analysis for structural safety, *Civil Engineering Systems,* **6**(4), 190–9.

Hauptmanns U. (1988) Fault tree analysis for process plants. In A. Kandel and E. Avni, *Engineering Risk and Hazard Assessment,* Vol. 1, CRC Press, Florida, 21–60.

Henley, E.J. and Kumamoto, H. (1981) *Reliability Engineering and Risk Assessment,* Prentice-Hall, Englewood Cliffs, New Jersey.

HSE (1978) *Canvey: An Investigation of Potential Hazards from Operations in the Canvey Island/Thurrock Area,* Health and Safety Executive, HMSO, London.

HSE (1990) *A Guide to the Control of Industrial Accident Hazard Regulations 1984,* HS(R)21 (Rev), Health and Safety Executive, HMSO, London.

IAEA (1992) *Procedures for Conducting Probabilistic Safety Assessments of Nuclear Power Plants (Level 1),* Safety Series No. 50-P-4, International Atomic Energy Agency, Vienna.

IEEE 352, *IEEE Guide for General Principles of Reliability Analysis of Nuclear Power Generating Station Protection Systems,* IEEE Std. 352-1975, Institute of Electrical and Electronic Engineers, New York.

Kalfsbeek, H.W. (1987) The OARS data bank: a tool for the assessment of safety related operating experience. In A. Amendola and A.Z. Keller (eds), *Reliability Data Bases,* D. Reidel Publishing Company, Holland, 317–42.

Kletz, T. (1986) *HAZOP and HAZAN – Notes on the Identification and Assessments of Hazards,* Institution of Chemical Engineers, England.

Lawley, H.G. (1974) Operability studies and hazard analysis, *Chemical Engineering Progress,* **70**(4), 45–56.

Lee, W.S., Grosh, D.L., Tillman, E.A. and Lie, C.H. (1985) Fault tree analysis, methods and applications – a review, *IEEE Transactions on Reliability,* **R-34**(3), 194–203.

Lees, F.P. (1980) *Loss Prevention in the Process Industries,* Vols 1 and 2, Butterworths, London.

Leveson, N.G. (1986) Software safety: why, what, and how, *ACM Computing Surveys,* **18**(2), 125–63.

Loss, J. and Kennett, E. (1987) *Identification of Performance Failures in Large Structures and Buildings,* School of Architecture and Architecture and Engineering Performance Information Center, University of Maryland.

Melchers, R.E. (1987) *Structural Reliability: Analysis and Prediction,* Ellis Horwood, Chichester, UK.

MIL-STD-1629, *Procedures for Performing a Failure Modes and Effects Analysis,* Military Standard MIL-STD-1629, Department of Defense, Washington, DC.

Montague, D.F. (1990) Process risk evaluation – what method to use?, *Reliability Engineering and System Safety,* **29**(1), 27–53.

O'Connor, P.D.T. (1991) *Practical Reliability Engineering,* John Wiley & Sons, Chichester, UK.

References

Recht, J.L. (1966) *Failure Modes and Effects Analysis*, National Safety Council.

RSS (1975) *Reactor Safety Study: An Assessment of Accident Risks in U.S. Commercial Nuclear Power Plants*, WASH-1400 (NUREG-75/014), US Nuclear Regulatory Commission, Washington, DC.

Simiu, E., Bietry, J. and Filliben, J.J. (1978) Sampling errors in estimation of extreme winds, *Journal of the Structural Division*, ASCE, **104**(ST3), 491–501.

Slater, D.H. and Cox, R.A. (1985) Methodologies for the analysis of safety and reliability problems in the offshore and gas industry. In F.Y. Yokel and E. Simiu (eds), *Application of Risk Analysis to Offshore Oil and Gas Operations – Proceedings of an International Workshop*, National Bureau of Standards Special Publication 695, US, 99–129.

Swain, A.D. and Guttman, H.E. (1983) *Handbook of Human Reliability Analysis with Emphasis on Nuclear Power Plant Applications*, NUREG/CR-1278, US Nuclear Regulatory Commission, Washington, DC.

Vesely, W.E. and Rasmuson, D.M. (1984) Uncertainties in nuclear probabilistic risk analysis, *Risk Analysis*, **4**(4), 313–22.

Villemeur, A. (1991) *Reliability, Availability, Maintainability and Safety Assessment*, John Wiley & Sons, Chichester, UK.

Virolainen, R. (1984) On common cause failures, statistical dependence and calculation of uncertainty; disagreement in interpretation of data, *Nuclear Engineering and Design*, **77**, 103–8.

CHAPTER 4

Performance of system elements

4.1 INTRODUCTION

Once the system has been modelled, a risk analysis can be performed using the model as the framework for the calculations. Such calculations imply a quantitative approach to risk analysis, and this requires quantitative descriptions (data) of the performance of the elements of the system. As discussed in section 3.5, such performances can be represented using random variables and probability theory. Typically the main variables of interest are:

1. reliabilities of components and items of equipment (e.g. pumps, valves, computers); and
2. resistances, strengths or capacities (e.g. structural, hydraulic, geotechnical, mechanical, electrical);
3. loads, demands or stresses (e.g. floods, waves, earthquakes);
4. undesirable consequences of failure.

As described in some detail in Chapter 3, probabilistic models for random variables typically are developed from statistical analyses of data. Such data is obtained from a variety of sources – these will be described in this chapter. Of course, the sources of data and the proposed probabilistic models based on that data are subject to several sources of uncertainty. It is important to recognize that uncertainties are also associated with the models for element performance. The uncertainties associated with the development and use of probabilistic models have been described in Chapter 3, and may be summarized as:

1. natural uncertainty of the probabilistic phenomenon itself;
2. modelling uncertainty; and
3. parameter uncertainty.

The present chapter will review the probabilistic models used in describing the performance of the elements or sub-systems (and which therefore influence the performance of the overall system). In addition, for some sub-systems the performance must be estimated primarily from models rather than directly from data. This is particularly the case for external effects such as wind, wave and earthquake loading. This important aspect is discussed in section 4.3. The treatment of uncertainty in probabilistic models is presented herein also. Finally, some comments are made about appropriate representation of consequences.

4.2 RELIABILITY AND FAILURE DATA

4.2.1 Description of data

The term 'reliability data' generally refers to the collected results of observations about the reliability of components and equipment. More properly the data mostly should be termed 'failure data' since it usually describes failure events rather than non-failure events (i.e. reliability). Provided this minor problem in terminology is understood, there should be no difficulty. Failure data is the data used for risk analysis, even though it itself is likely to be the collective results of earlier experiments. Typically such data is given in terms of

1. a failure (or hazard) rate (e.g. 0.01 faults/year), or
2. a mean time to failure.

The failure rate is the number of failures expected during a given time period (usually the time period during which the element is actually in operation or use). It is the measure of component (or subsystem) reliability most often used in risk analyses. The mean time between failures or mean time to first failure are other terms used (see Chapter 6).

The way 'failure' is defined depends on the element or component and its function. Typically, component performance is considered in terms of discrete possible outcomes; such as success or failure. For components operating continuously failure may be defined simply as failure during operation. For components on standby or operating intermittently, failure may be defined additionally as

1. failure to operate on demand,
2. operation before demand, or
3. operation after demand has ceased.

The discussion to follow will deal only with reliability data as it relates to equipment malfunctions caused by internal events (normal wear and tear of components) or by some external events (earthquake, fire). Failures due to human error (e.g. maintenance or test personnel error) should be collected and analysed separately in order that estimates for human reliability can be calculated for that task (see Chapter 5).

4.2.2 Types of reliability and failure data

The data used for risk analysis depends to some degree on the nature of the system. At least one or more of the following reliability data are needed for all risk analyses:

1. overall failure rates;
2. failure rates in individual failure modes;
3. variation of failure rates with time;
4. unavailability on demand; and
5. repair times.

These different aspects are addressed below.

4.2.3 Overall failure rates

Often the most convenient (and most accurate) way to obtain estimates of the reliability of components is by collecting and statistically analysing appropriate failure data. For example, failure data may be obtained from operating experience if the time, type and cause of failures are recorded over an appropriate period of time. Reliability data can then be calculated; this data is typically presented in the form of 'reliability databases'. Reliability databases may therefore be conveniently used for future reference by other users. The collection, analysis and use of failure and reliability data is described in section 4.3.

Alternatively, reliability data may be obtained from the physical understanding of the behaviour of the components or sub-systems under failure conditions. In this case the probability of failure is calculated directly from reliability theory dealing with 'load-resistance' (or 'demand-capacity') sub-systems. This approach assumes that the component or subsystem fails if the demand or the load exceeds the capacity or strength. It is more complex than the use of simple observationally derived failure rates and has not found much application in electrical or mechanical systems. It is used principally for predicting the reliability of unique or expensive components or items of equipment, or where environmental or external loads are involved. The theory for it has been developed principally in structural engineering, but the basics were outlined already much earlier, including in electrical engineering texts dealing with overall systems reliability (Shooman, 1968). The basic theoretical ideas are introduced in section 4.4. As will be seen, the theory requires information about the probability distributions describing the resistance (capacity, strength) and the load (demand, stress). This type of information is decribed in some detail for various external loadings and for various resistances or capacities in section 4.4.

4.2.4 Failure rates in individual failure modes

The reliability of a component may be influenced by one or more of the following factors:

1. definition of failure;
2. complexity and maturity of manufacturing process;
3. process or operating environment (humidity, pressure, temperature, dust);
4. operating stress (load placed on component – power, voltage, current); and
5. maintenance practices.

It follows that reliabilities of components and sub-systems are generally plant or site specific and hence related to the failure modes specific to the system. It would be expected, therefore, that failure rates even for similar components might vary from system to system. This is likely to be the case for chemical and petro-chemical facilities or plants, since operational conditions can vary widely. Generic data, such as obtained from failure or reliability databanks might now be misleading. More focused and more specific data is now required for high-quality risk analysis. This may involve experiments on the plant or components of particular interest.

4.2.5 Variation of failure rates

It is reasonable to expect that failure rates will vary with time, but quantitative information on this phenomenon is limited. It has frequently been suggested that the relationship between failure rates and time may be represented by the 'bathtub' curve (e.g. Daniels, 1982), see Figure 4.1. It shows that very early in the life of a typical component the failure rate reduces sharply with time. This is sometimes referred to as the 'burn-in' phase for mechanical and electrical components. In other situations, such as with road bridges, a similar pattern has been noted, with many construction failures and failures immediately after construction, followed by a period of rather few failures. Of course, such results are based on statistics for many components or systems – the bathtub curve being merely an average trend. As the component or system ages, fatigue, corrosion and other processes lead to an increase in the failure rate – the so-called 'burn-out' phase in electrical engineering risk, or the 'wear-out' phase in some other branches of engineering.

The validity of the bathtub curve to all types of components has been questioned widely and cannot be considered always to be valid (e.g. Lees, 1980). For example, it has been suggested that in the aviation industry only 4% of components conform to the bathtub curve. Instead, it was found that the most likely pattern was high infant mortality

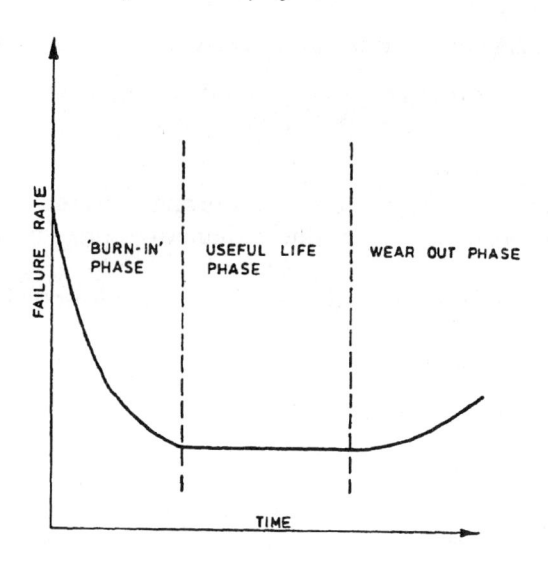

Figure 4.1 Typical 'bathtub' curve.
Source: Daniels (1982).

followed by random failure with no wear-out zone (Kannegieter, 1993). Nonetheless, it is generally assumed that testing procedures and monitoring of new equipment will detect most of the failures occurring during the 'burn-in' phase, and that 'wear-out' failures are reduced by appropriate maintenance procedures. For these reasons and for ease of calculations (see Chapter 6), it is often assumed in risk analyses that failure rates are constant throughout the life of the component. Note that for software, however, it is more likely that the failure rate will decrease over time as more faults (or bugs) are detected and corrected (e.g. Hecht and Hecht, 1986). The effect of time on failure rates (and system evaluation) is described in more detail in section 6.7.

4.2.6 Unavailability

The unavailability of a component is an important reliability measure for standby systems. (A standby (sub-)system is one which is expected to become operational if the system itself fails.) Unavailability refers to the standby component or system being out of action or for some other reason not being able to take over the functions of the (usual or main) system. Typically, the time to failure of a component during a standby period is modelled as an exponentially distributed random variable, in the same way as used for normal components (e.g. IAEA, 1992). The time-dependent unavailability may be modelled as an average (or constant) value, based, for example, on the frequency of testing or other

Sources of failure data 83

influences related to unavailability to meet demand. In such cases, unavailability is conveniently treated as a constant failure rate.

4.2.7 Repair times

Repair times may influence the ability of the system to recover from one or more component failures; for instance, a system may deteriorate while a critical component is in a 'failed' state.

4.3 SOURCES OF FAILURE DATA

In practice, failure data for components and equipment may be obtained from four principal data sources (e.g. Bello, 1987):

1. laboratory testing;
2. field data collection;
3. incident data banks; and
4. expert opinions.

The first three sources of failure data are essentially the same and may be categorized collectively as 'experimental data'. The only difference lies in the way each is available to the analyst.

4.3.1 Experimental data

Reliability data may be obtained from a statistical analysis of experimental failure data. The reliability measure of most interest to the risk analysts is the average failure rate $\lambda(t)$ for a time interval $(0 - t)$; this is typically calculated as

$$\lambda(t) = \frac{n}{t_L S} \qquad (4.1)$$

where n is the number of component failures observed in a given time interval (t_L – calendar or operational time), and S is the number of components at the start of the given time interval. Therefore, experimental data for the parameters n, t_L and S needs to be collected. This data may be obtained from laboratory testing, field data or incident data banks.

In laboratory testing, failure events and time intervals (i.e. n, t_L) may be recorded for similar components tested to destruction (i.e. S = 1). However, care must be taken to ensure that the 'in-service' environment and operating conditions of the components are simulated properly in the laboratory test. Failure data for electrical and electronic components are generally obtained from laboratory testing (e.g. Colombo, 1987; Verwoerd, 1991), probably because these components are manufactured on a large scale, have small unit costs and so can be tested in large numbers. Laboratory testing is often too expensive and thus not appropriate for mechanical components.

Field data comprises failure event and time interval data obtained from operating and maintenance records of the system. The collected failure data may be aggregated for (1) a single plant or site (i.e. plant or site specific failure data) or (2) a number of plants (i.e. general or generic failure data for a population of plants). It is highly desirable that field data be collected in a consistent and systematic manner, as this will allow reliability data to be calculated with more confidence. Note that if failure events are rare then the uncertainty associated with the calculated reliability data may be unacceptably high.

Alternatively, failure data for a risk analysis can be obtained from incident databanks; these generally contain information on failure events, their cause, consequences and time of occurrence, see section 3.3.5. From databank information, generic reliability data may be estimated. Unfortunately, the techniques used and the assumptions made in the collection and recording of databank information are not always presented and hence cannot be assessed by the analyst. Hence, the analyst often has to make a subjective assessment as to the appropriateness or validity of the information for the specific component under consideration. Most incident databanks are developed for use by a single industry or engineering system, and failure data is collected from a number of sites or plants. For this reason, reliability data calculated from incident databank failure data is normally generic in nature and is not plant specific (see also Henley and Kumamoto (1981) and Daniels (1982) for an overview of several databanks).

There are some uncertainties associated with the collection and analysis of failure data when the data is obtained from laboratory testing, field data and incident databanks. This may be due to one or more of the following deficiencies in the failure data (Duphily and Long, 1977):

1. non-standardized data reporting;
2. poor descriptions of the item in question;
3. time-consuming data feedback;
4. lack of accuracy in repair times required for many parts;
5. inconsistent and vague definitions of terms on the reporting forms;
6. difficulty in pinpointing the cause(s) of the failure;
7. government security or company proprietary classification of the data; and
8. lack of data on the relationship between the part's failure and its operating environment.

Because of their importance, these uncertainties will now be described in a little more detail. Failure data records tend to describe 'malfunction' events that were identified during the operation or maintenance of a system. A major source of uncertainty, therefore, is the selection of which events out of all the malfunction events constitute 'failure'. Mosleh (1986) has defined failure as the event in which 'a piece of equipment fails to perform a function required by the system model'. For example, the event 'pump fails to operate on demand' should be considered a component failure if the system model is 'the operation of a nuclear power plant', but it may be of little importance if such an event occurs in a different system. Therefore, the severity of the malfunction tends to be influential in defining 'failure'. Evidently, often subjective judgement is required. For example, a malfunction event may be 'pump shaft vibrations' that were identified (and repaired) by maintenance staff. The analyst must decide whether the vibrations were severe enough to cause imminent shaft seizure if the pumps were not repaired at that time. Naturally, this decision is made more difficult if the records are unclear or incomplete. Uncertainties in the definition of failure may not be particularly important for components with relatively high failure rates. However, if only a few failures are identified then uncertainty in the calculated failure rate may be very high. For this reason, Mosleh (1986) suggests that analysts tend to be conservative. In other words, a component is deemed to have 'failed' if the analyst is unsure of its classification.

The collection of exposure data (such as the cumulative hours of operation of a component) is also subject to some uncertainty, but the range of uncertainty is generally less than that associated with estimating the number of failure events (Mosleh, 1986). For some components, accurate exposure data may be obtained directly from operation time meters installed on the equipment. However, for most components exposure data is estimated from plant records. These documents describe test procedures, component functions, operational times and outages. Evidently, the use of this data for reliability estimates requires subjective judgements and perhaps some assumptions. For example, run times during tests must be extrapolated to provide exposure data for normal operating conditions and extrapolated even further for estimates under extreme conditions. Neither extrapolation may not be particularly accurate because many components do not operate continuously, but actually operate intermittently or on demand. Mosleh (1986) cites a case in which exposure times obtained from equipment operation time meters were 300% greater than exposure times estimated from test procedures. (Note that exposure data may also be measured in terms of number of cycles or number of demands.)

The expression for failure rate given by Equation (4.1) is only accurate if the component is new at the start of the test time interval, and a failure coincides with the end of the test time interval (or 'observation window'). In an operational system, it is often the case that these conditions are not met (except for laboratory testing), as shown in Figure 4.2. Obviously, it will be conservative simply to assume that each component was just

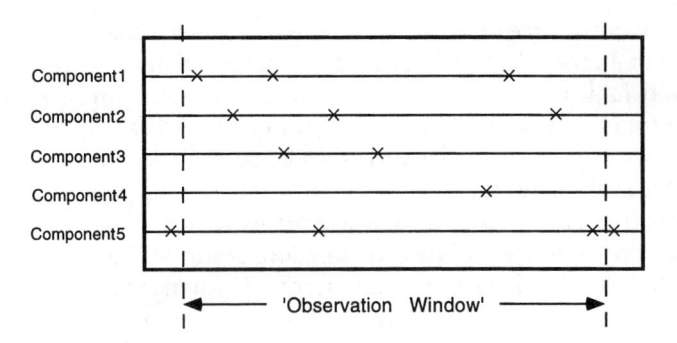

```
                  |                                             |
Component1  |—×————————×——————————————×————|
            |                                             |
Component2  |————×—————————×———————————————×——|
            |                                             |
Component3  |————————×————————×————————————————|
Component4  |———————————————————————×———————|
Component5  |×—————————×————————————————————×|×
            |                                             |
            |←———————— 'Observation Window' ————————→|
```

Figure 4.2 Observation of failures in an operational system (x = failure).

about to fail at the end of the test time interval. This assumption is also convenient in the case where there are no observed failures at all in the test period. Several alternative approaches for calculating failure rates under these conditions have been suggested, including the assumption of independent events and the use of the binomial distribution for estimating the failure rate, however, Lees (1980) concludes 'that there is no entirely satisfactory method of dealing with this case'.

The inherent variability of seemingly identical components, as well as uncertainties in failure and reliability data, indicate that failure rates for identical components should be treated as random variables, as discussed in section 3.5. As a probabilistic description of the random variable, a lognormal distribution for individual component failure rate is often assumed; however, other distributions also may be used (e.g. Mosleh, 1986; Colombari, 1987). The spread of the probability distribution of individual failure rates may be expressed as error bounds, assessed range, uncertainty range or confidence limits. For example, RSS (1975) and IEEE (1984) provide failure rates for the 'best estimate', and low/minimum and high/maximum values (typically 5th and 95th percentiles respectively). The 'best estimate' often is assumed to be the median of the lognormal distribution of component failure rates as this is a more conservative measure than the mean (e.g. Apostolakis *et al.*, 1980). Note that a probability distribution of individual failure rates can not be obtained solely from experimental failure data as such data tends to reflect mainly the 'tail' of the distribution. Also note that a confidence interval for the average failure rate represents only a measure of the uncertainty in the average failure rate, and does not represent the inherent (and natural) variability to be expected in identical components.

4.3.2 Expert opinion

Qualitative and quantitative information obtained from operators, maintenance staff and management (or others), if of sufficient quality, may provide useful failure and/or reliability data. Such data relies, of course, heavily on the experience and knowledge of the people consulted and their ability to make judgements and to convey opinions. Typically, they may be asked to estimate the average failure rate and/or to estimate the range of failure rates. From this information it may be possible to derive point estimates or a probability distribution for individual failure rates. Krinitzsky (1993), in dealing with seismic risk assessments, asserts that some experts may be 'fee-hungry knaves, time servers, dodderers in their dotage ... Yet, these and all sorts of other characters can pass inspections, especially when their most serious deficiencies are submerged in tepid douches of banality.' Evidently, great care must be taken in the selection of experts, and at least some of their estimates ought to be calibrated against known component (or other) reliability data. Several techniques exist to assist in the process of converting qualitative and quantitative information into reliability data; these include Absolute Probability Judgement (using the Delphi technique); Paired Comparisons; or the Expert Information approach (e.g. Daniels, 1982; Kaplan, 1992). These methods essentially aggregate expert opinions so that a relatively high degree of 'consensus' or consistency between experts is attained.

Green (1983) has summarized the results of several studies in which comparisons were made between failure rates of electronic and mechanical components obtained from expert opinions and those obtained from existing reliability data (i.e. calculated from extensive field data). It was found that at least a third of the experts' opinions were within a factor of three of the field data or the recognized reliability data. It was also observed that no expert provided consistently good estimates. Nonetheless, it was concluded that generally there was good agreement between the estimated failure rates and the field data or recognized failure rates.

However, Apostolakis (1986), Martz (1984), and others have asserted that there are significant uncertainties associated with the analysis and use of reliability data elicited directly from experts. For example, 12 expert estimates of average pipe failure (rupture) rates were obtained in the study reported by RSS (1975), see Table 4.1. Note that the expert estimates given in Table 4.1 show a significant degree of expert to expert variability. The reliability data recommended that the median estimate be 1×10^{-10} failures per hour per pipeline section, with 5th and 95th percentiles (i.e. assessed range) of 3×10^{-12} and 3×10^{-9} respectively (RSS, 1975). However, Table 4.1 shows that eight expert estimates were greater than the upper bound; this raises the question: 'why were these expert estimates ignored?'. According to Apostolakis (1986), this question would have been addressed by the RSS (1975) working group, but suggests that 'this case shows how significant the RSS (1975) group's judgement could be in the determination of the failure rate distribution'.

Table 4.1 Expert estimates of pipe failure rates

Source no.	Expert estimate	Source no.	Expert estimate
1	5×10^{-6}	7	2×10^{-10}
2	2×10^{-9}	8	3×10^{-9}
3	1×10^{-10}	9	1×10^{-8}
4	1×10^{-6}	10	1×10^{-8}
5	7×10^{-8}	11	6×10^{-9}
6	1×10^{-8}	12	1×10^{-8}

Source: adapted from RSS (1975).

In view of these observations, it is not surprising that generic failure rate distributions have been considered by some to be too narrow, and that they should be broadened to represent less confidence in the expert estimates. One such approach is to broaden the failure rate distribution by redefining the endpoints of the assessed range so that they represent the 20th and 80th percentiles (Apostolakis et al., 1980; Apostolakis, 1982), however, this may be a somewhat conservative approach (Martz, 1984).

A further source of uncertainty occurs in expert opinions because 'it appears that people become very sceptical when probabilities of failure smaller than 10^{-5} per demand are reported, when the numbers are not supported by any statistical evidence' (Apostolakis, 1986). For this reason, the assessed range is generally largest for events with very low probabilities of occurrence (Vesely and Rasmuson, 1984). Despite these and other problems, Paté-Cornell (1986) concludes that 'experts' opinions are indispensable given the scarcity of unquestionable data sets'.

4.3.3 Bayes Theorem – combining different data

In some cases, only a limited amount of experimental failure data may have been collected (e.g. due to insufficient operational experience) or the failure data collected may not have contained sufficient information on rare failure events. It may be desirable also to represent individual failure rates by a probability distribution. In these cases, the data may be combined with other relevant data, such as generic data or expert opinion, using Bayes Theorem. Typically this will involve:

1. experimental failure data obtained for a specific plant or situation, and
2. reliability data obtained from (generic) reliability databases or expert opinions.

In using Bayes Theorem, the approach is to commence with a so-called 'prior' distribution. In the present case, the prior distribution is that which represents the generic reliability data (2). Failure data information is now considered to represent 'new evidence' and is used to 'update' the prior distribution to produce a 'posterior' distribution that incorporates the new evidence or knowledge. Typically, the new evidence is experimental data. If it is given in terms of the number of failures out of S trials, it may be represented by the Binomial distribution; if it is given as the number of failures during a specific operating time it may be represented by the Poisson distribution. In each case these distributions represent the 'likelihood function'. The posterior distribution is calculated from

$$f''(\lambda) = \frac{f'(\lambda) L(E \mid \lambda)}{\int_0^\infty f'(\lambda) L(E \mid \lambda) d\lambda} \qquad (4.2)$$

where $f'(\lambda)$ is the prior distribution, $L(E \mid \lambda)$ is the likelihood function or conditional probability of observing the experimental outcome E given that the value of the parameter is λ, and λ is the failure rate (e.g. Ang and Tang, 1975). If the experimental data is the number of failures (n) during a specific operating time (t_L), then the likelihood function is the Poisson distribution given by

$$L(E \mid \lambda) = \frac{\exp(-\lambda T_L)(-\lambda T_L)^n}{n!} \qquad (4.3)$$

Typical prior, likelihood and posterior distributions are shown in Figure 4.3, for the example of the failure of a diesel generator while running. In this example, the prior distribution was assumed to be a lognormal distribution with 5th and 95th percentiles given by 3×10^{-4} and 3×10^{-2} (per hour) respectively, obtained from a generic reliability database (RSS, 1975). The observational data (the new evidence) for the plant was taken as nine failures (n) during 398.03 hours of operation – this provides t_L. The posterior distribution was then estimated from Equation (4.2) using the likelihood function given by Equation (4.3). Note that the new evidence shifted the posterior distribution towards higher failure rate values.

In many cases the posterior distribution is highly dependent on the selection of the prior distribution. Hence considerable care is required in its selection. It has also led to a warning that the Bayesian approach should not be used to provide a justification for reduced experimental testing (e.g. O'Connor, 1991). Direct testing of a component is still the best way of obtaining reliable data.

The above discussion of Bayes Theorem has been in terms of reliability data. However, it is applicable to any situation in which existing probabilistic knowledge is to be updated with new evidence. For risk analyses this can include updating of probabilistic information on strengths, loads, consequences and other system elements.

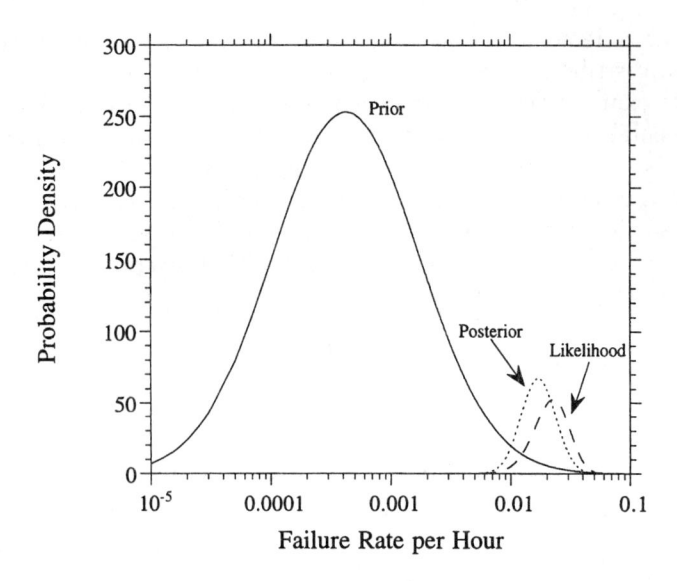

Figure 4.3 Prior, likelihood and posterior distributions for diesel generators – failure while running.
Source: data from Apostolakis *et al.* (1980).

4.3.4 Reliability databases

Reliability databases contain reliability data obtained from

1. statistical analysis of failure data; and
2. reviews of reliability data reported in existing literature (e.g. journals, conference proceedings, research reports).

These reliability databases may be produced in printed or computerized formats. Reliability data obtained from reliability databases is normally used directly in a risk analysis if failure data has not been collected for the particular plant or system of interest, if the system is not yet operational, or as a prior distribution if experimental failure data is to be used to produce a posterior distribution using Bayes Theorem. On the other hand, if large amounts of failure data have been collected for the system of interest, then reliability data may be obtained directly from a statistical analysis of failure data. In this case reference to a reliability database may not be necessary, except to compare and validate the calculated reliability data with reliability data obtained from other studies.

A list of some typical component reliability databases is given in Table 4.2. A large proportion of the databases have been developed for use in risk analyses for the operation of nuclear power plants. However, databases have more recently been developed for other applications; namely, chemical process plants, telecommunication systems and

Table 4.2 Component Reliability Databases.

Components or systems	Source
Nuclear power plants	
International	IAEA (1988)
United States	RSS (1975), IEEE (1984)
Europe (CEDB)	Balestreri (1987)
United Kingdom (SYSREL)	Daniels (1982)
Electronic	DOD (1986)
Telecommunications	Garnier (1987)
	Pirovano and Turconi (1988)
Chemical process plants	CCPS (1989)
Offshore platforms (OREDA)	OREDA (1984)
LNG facilities	Johnson and Welker (1981)
Oil and gas pipelines	Anderson and Misund (1983)

offshore platforms. Fortunately, components and other items of equipment such as switches, pumps and electric motors are sometimes common to several engineering systems; hence, some reliability databases are directly applicable to more than one engineering system. Note that many reliability databases are continually being updated with new failure data. For example, by 1982 the SYSREL reliability database contained 11 000 entries of summarized reliability data (with a total operating experience of 12×10^{12} hours), and in that year approximately 1500 new entries were added (Daniels, 1982). It follows, therefore, that some reliability databases contain reliability data based on large amounts of operating experience by combining failure data from more than one plant.

Reviews of reliability databases are available for:

1. offshore platforms (Verwoerd, 1991);
2. nuclear power plants (Apostolakis *et al.*, 1980);
3. chemical and oil process and storage plants (Lees, 1980);
4. industrial plants (McFadden, 1990); and
5. engineering systems in general (Henley and Kumamoto, 1981; Dhillon, 1988).

Reliability data obtained from several reliability databases are shown in Tables 4.3 to 4.6. For each component, the reliability is described by the following statistical parameters: average (or 'best estimate') failure rate; and/or low/minimum and high/maximum values (i.e. error bounds). Some databases also provide information on the influence of manufacturing processes, and operating environments and stresses (e.g. IEEE, 1984; DOD, 1986), see Table 4.4.

Most existing reliability databases contain statistical parameters for generic reliabilities; this refers to failure data obtained from a number of

Table 4.3 Reliability data for temperature instruments, controls and sensors

Failure mode	Failures/10^6 hours			Failures/10^6 cycles			Repair time (hours)		
	Low	Rec	High	Low	Rec	High	Low	Rec	High
All Modes	0.31	1.71	21.94	0.11	0.75	1.51	0.3	0.74	1.3
Catastrophic	0.13	0.70	9.0						
Zero or maximum output	0.06	0.31	4.05						
No change of output with change of input	0.01	0.04	0.45						
Functioned without signal	0.03	0.18	2.34						
No function with signal	0.03	0.17	2.16						
Degraded	0.14	0.75	9.65						
Erratic output	0.03	0.17	2.22						
High output	0.03	0.15	1.93						
Low output	0.01	0.06	0.77						
Functioned at improper signal level	0.05	0.29	3.67						
Intermittent operation	0.02	0.08	1.06						
Incipient	0.04	0.26	3.29						

Note: Rec refers to the 'Best Estimate'.
Low, High refers to the best and worst data points (i.e. this establishes the range).
Source: adapted from IEEE (1984).

Table 4.4 Environmental factors for temperature instrument, control and sensor reliability data

Environmental stress	Modifier for failure rate
High temperature	× 1.75
High radiation	× 1.25
High humidity	× 1.5
High vibration	× 2.0

Source: adapted from IEEE (1984).

Table 4.5 Reliability data for mechanical and electrical components

Component and failure mode	'Best estimate'	Upper and lower bounds
Electric motors:		
Failure to start	3×10^{-4}/D	$1 \times 10^{-4} - 1 \times 10^{-3}$
Failure to run (normal)	1×10^{-5}/hr	$3 \times 10^{-6} - 3 \times 10^{-5}$
Failure to run (extreme environment)	1×10^{-3}/hr	$1 \times 10^{-4} - 1 \times 10^{-2}$
Battery power systems:		
Failure to provide proper output	3×10^{-6}/hr	$1 \times 10^{-6} - 1 \times 10^{-5}$
Switches:		
Limit – failure to operate	3×10^{-4}/D	$1 \times 10^{-4} - 1 \times 10^{-3}$
Torque – failure to operate	1×10^{-4}/D	$3 \times 10^{-5} - 3 \times 10^{-4}$
Pressure – failure to operate	1×10^{-4}/D	$3 \times 10^{-5} - 3 \times 10^{-4}$
Manual – fail to transfer	1×10^{-5}/D	$3 \times 10^{-6} - 3 \times 10^{-5}$
Contacts short	1×10^{-7}/hr	$1 \times 10^{-8} - 1 \times 10^{-6}$
Pumps:		
Failure to start (normal)	1×10^{-3}/D	$3 \times 10^{-4} - 3 \times 10^{-3}$
Failure to run (normal)	3×10^{-5}/hr	$3 \times 10^{-6} - 3 \times 10^{-4}$
Failure to run (extreme environment)	1×10^{-3}/hr	$1 \times 10^{-4} - 1 \times 10^{-2}$
Valves (motor-operated):		
Fails to operate	1×10^{-3}/D	$3 \times 10^{-4} - 3 \times 10^{-3}$
Failure to remain open	1×10^{-4}/D	$3 \times 10^{-5} - 3 \times 10^{-4}$
External leak or rupture	1×10^{-8}/hr	$1 \times 10^{-9} - 1 \times 10^{-7}$
Circuit breakers:		
Failure to operate	1×10^{-3}/D	$3 \times 10^{-4} - 3 \times 10^{-3}$
Premature transfer	1×10^{-6}/hr	$3 \times 10^{-7} - 3 \times 10^{-6}$
Fuses:		
Premature, open	1×10^{-6}/hr	$3 \times 10^{-7} - 3 \times 10^{-6}$
Failure to open	1×10^{-5}/D	$3 \times 10^{-6} - 3 \times 10^{-5}$
Pipes:		
< 75mm, rupture	1×10^{-9}/hr	$3 \times 10^{-11} - 3 \times 10^{-8}$
> 75mm, rupture	1×10^{-10}/hr	$3 \times 10^{-12} - 3 \times 10^{-9}$
Welds:		
Leak, containment quality	3×10^{-9}/hr	$1 \times 10^{-10} - 1 \times 10^{-7}$

Source: adapted from RSS (1975).

Table 4.6 Reliability data for fire water pumps on offshore platforms

Population	Samples	Aggregated time in service (10^6 hrs)		Number of demands
17	10	Calendar time .3826	Operational time .0002	1135

Failure mode	No. of Failures	Failure rate (per 10^6 hrs)			Repair (manhours)		
		Lower	Mean	Upper	Min.	Mean	Max.
Critical							
	80*	120	210	310	–	86	–
Failed to start	13†	26000	47000	78000			
	75*	100	190	90	24	86	120
	9†	6200	32000	69000			
Failed while running	5*	2.0	23	51	3	93	130
	4†	4600	15000	36000			
Degraded							
	24*	30	71	120	–	180	–
	3†	0	14000	45000			
High temperature	22*	22	66	120	6	190	400
	3†	0	14000	44000			
Low output	1*	0.14	2.6	12	–	–	–
Unknown	1*	0.14	2.6	12	–	96	–
Incipient							
Unknown							
All Modes	303*	680	840	1000	–	81	–
	45†	87000	180000	280000			

Note: *denotes calendar time
†denotes operational time
Source: adapted from OREDA (1984).

different sources. Therefore, reliability data obtained from a single reliability database may be applied to different plants or even to different systems, such as nuclear, chemical and offshore industries. Hence, generic reliability data for a specific component includes within it the variability due to different manufacturers, models, operating and maintenance procedures, systems, and the inherent (and natural) variability to be expected in identical components. Further, it may be unclear if the recommended probability distributions refer to the distribution of average failure rates or the distribution of the population of individual failure rates. Therefore, generic data often does not come from exactly the same population, and the statistical aggregation (or pooling) of this data may result in unrealistically low error bounds due to the large sample size. This variability is often associated with average failure rates, and so is not representative of variability to be expected in a single plant or site.

It is therefore apparent that uncertainties exist in the use of reliability databases that contain generic reliability data. Mosleh (1986) suggests that this uncertainty is mainly due to the fact that some reliability databases fail to specify:

1. precise nature of the component;
2. definition of failure;
3. source(s) of failure data;
4. definition of 'best estimate' (either mean or median);
5. whether error bounds relate to variability between plants or variability within a plant; and
6. type of probability distribution to use.

If a reliability database contains any of these deficiencies then the analyst may be unsure as to the validity of the generic reliability data, and if valid, how the data should be used.

Not surprisingly, the uncertainties associated with the collection, analysis and use of failure and reliability data have led to the suggestion that 'reliability engineers are accustomed to working with failure data which predicts behaviour within a factor of ten' (Henley and Kumamoto, 1981). Others have observed that there are order of magnitude differences in component failure rates among the various sources of reliability data used in nuclear power plant risk studies (Tomic and Lederman, 1989).

4.3.5 Influence of external factors

In a sense the reliability data described in the previous sections is concerned mainly with component failures caused by internal factors (e.g. metal fatigue, corrosion, wear, age). However, the influence of external factors on equipment performance is also an important consideration. Ground motions caused by an earthquake may result in damage to piping, motors or sensitive electronic safety systems. Other external factors of relevance to equipment performance include the occurrence of fires, aircraft crashes, flooding and lightning strikes (e.g. USNRC, 1989). The modelling of the effects of all these external influences is a matter for specialist disciplines and is beyond the scope of the present chapter. The effect of earthquakes on equipment reliability is discussed briefly below: this provides an example also of some of the uncertainties which may be associated with external factors or events.

A so-called 'fragility' curve is in essence a cumulative distribution function which represents the conditional probability of failure (for a component, or a piece of equipment, or a structure) as a function of an action or a force or other influence or demand. A typical fragility curve is shown in Figure 4.4. For seismic conditions, the fragility curve is often expressed as a function of peak ground acceleration (PGA), the measure used for estimating the response or failure rate for the component etc. under consideration. It is often taken as described by the lognormal distribution, as suggested in Figure 4.4.

The capacity or resistance of a component against PGA is given by the product of a design PGA – obtained from structural design codes or selected from past experience – and a factor of safety. The factor of

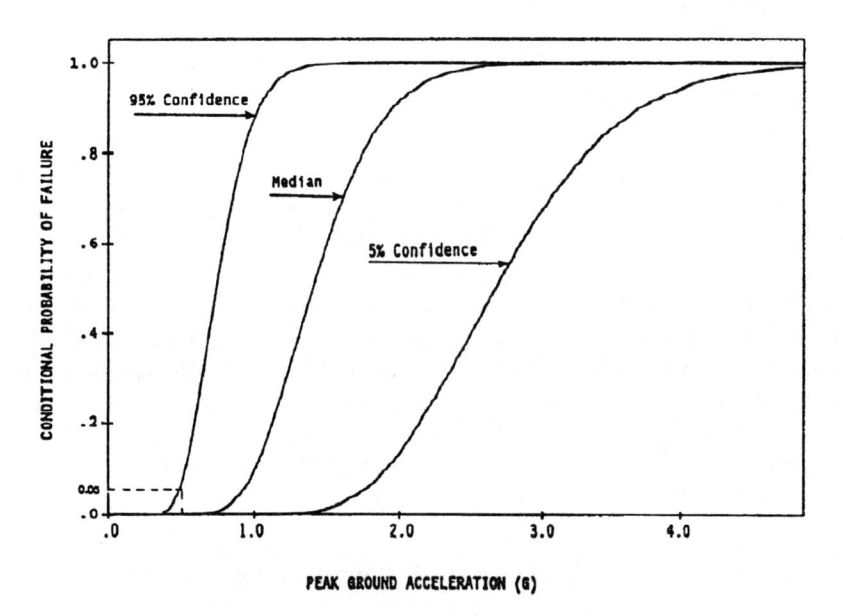

Figure 4.4 Typical seismic fragility curve.
Source: Ravindra *et al.* (1990).
Reprinted from M.K. Ravindra, M.P. Bohn, D.L. Moore and R.C. Murray (1990) Recent PRA applications, *Nuclear Engineering Design*, **123**, 155-66, with kind permission from Elsevier Science S.A., PO Box 564, 1001 Lausanne, Switzerland.

safety itself usually is taken as a product of a number of variables representing uncertainty and variability in

1. strength,
2. inelastic energy absorption,
3. spectral shape,
4. soil structure interaction,
5. modelling,
6. method of analysis/testing,
7. combination of failure modes, and
8. combination of earthquake components (Ravindra et al., 1990).

For each variable, the median and the logarithmic standard deviations representing randomness in uncertainty and variability in the median may be estimated from experimental data and/or expert opinions. Three statistical parameters can then be calculated to describe the PGA capacity of the component etc.: (1) median ground acceleration capacity (A_m); (2) standard deviation representing randomness in peak ground acceleration capacity (β_R); and (3) standard deviation representing uncertainty in A_m (β_U). Therefore, the median seismic fragility curve is described by A_m and β_R. The uncertainty in the PGA distribution can be represented by a confidence limit, such as the 5% confidence curve, obtained by using a reduced median capacity (obtained from the 5th percentile of the lognormal distribution with A_m and β_U as parameters) and β_R as parameters. Statistical parameters used for typical seismic fragility curves are given in Table 4.7, for equipment and structures found in nuclear power plants.

4.4 RELIABILITY OF LOAD-RESISTANCE SUB-SYSTEMS

Structural, geotechnical, mechanical, hydraulic, electrical, electronic and other systems are often represented as 'load-resistance' or 'demand-capacity' system elements. In many cases, these elements represent

Table 4.7 Statistical parameters for seismic fragility curves

Component/equipment/structure	A_m (g)	β_R	β_U
Aux. feedwater tank	0.82	0.17	0.42
Diesel gen. fuel DT	1.50	0.30	0.50
Dgn. rm. vent fans	1.30	0.30	0.40
Ceramic insulator failure	0.20	0.20	0.25
Oil cooler anchor bolt failure	0.91	0.24	0.43
Control building collapse	1.00	0.24	0.33
Cable trays	2.70	0.48	0.42

Source: adapted from Ravindra (1990); Ravindra, et al. (1990).

substantial and expensive sub-systems such as containment vessels, dams and other unique or 'one-off' elements. Naturally, their reliabilities cannot be directly inferred from observation of failures or other experimental studies. In these circumstances, reliabilities need to be predicted from predictive models and probabilistic methods. It is observed that many load-resistance systems occur in civil engineering systems, thus modelling efforts have been concentrated in this discipline. However, as will be discussed in the present section, computational and probabilistic methods derived for civil engineering systems are also appropriate for other systems, a matter foreshadowed already by Shooman (1968).

Loads may refer to demand, applied stress, applied voltage, internal stresses due to temperature changes, or structural loads (e.g. wind and dead loads). Conversely, systems are designed so that these and other loads are resisted by the system (or by a subsystem); this resistance is generally a physical property such as strength, capacity, adhesion, hardness or melting point. Examples are:

1. a transistor gate in an integrated circuit fails when the voltage applied causes a local current density, hence the temperature rises above the melting point of the semi-conductor material (O'Connor, 1991);
2. a vessel breach is induced if the containment pressure due to heating and high-pressure melt ejection exceeds the containment strength (IAEA, 1992);
3. a structural beam fails if bending actions due to dead and live loads exceed the bending capacity of the beam; and
4. a gravity dam fails if its resistance to sliding (at its foundation) is less than the flood and earthquake lateral loads (Bury and Kreuzer, 1985).

Failure is thus deemed to occur when the demand or load (Q) exceeds the capacity or resistance (R). This approach is sometimes referred to as 'load-resistance system', 'stress-strength', 'load-capacity', 'R-Q', 'load-strength interference', 'integral approach' or 'interaction diagrams' relationships (Klaassen and Peppen, 1989; Melchers, 1987; O'Connor, 1991; Davidson, 1988; IAEA, 1992). The probability of failure is, evidently:

$$p_f = \Pr(R \leq Q) = \Pr(R - Q \leq 0) = \Pr(G(R,Q) \leq 0) \qquad (4.4)$$

where $G()$ is termed the 'limit state function', in the present case this is equal to $R - Q$. Thus the probability of failure is the probability of exceeding the limit state function. Note that R and Q must be in the same units. In general, the limit state function may contain more than two variables, for instance, the limit state for hull girder bending can be expressed as

$$G(R,Q) = F_y Z - M_{sw} - M_w \qquad (4.5)$$

where the product $F_y Z$ represents the resistance of the hull girder, and M_{sw} and M_w are the still-water and wave-induced load effects (bending moments) respectively (White et al., 1995).

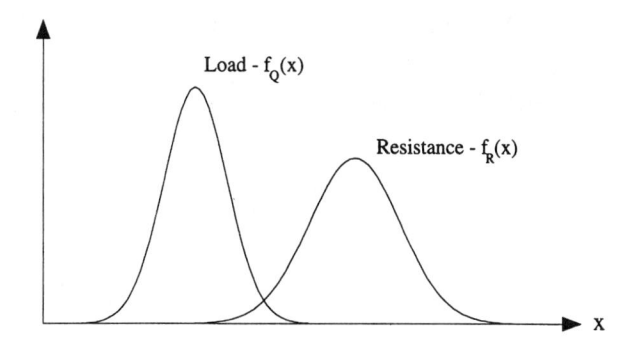

Figure 4.5 Loads and resistances as random variables.

Since, the strength or capacity of identical components will vary from component to component due to variabilities in material properties, production processes, geometric dimensions, environmental conditions, maintenance, etc. it is appropriate to represent strengths or capacities as random variables (see section 3.5). Similarly, loadings or demands are influenced by a variety of factors, such as environment, temperature, geographic location (e.g. wind). Some are also time dependent and highly variable, such as wind velocities. Consequently, loads or capacities also should be modelled as random variables. In the case of Equation (4.4) with one load (Q) and one resistance (R), the relation between them is as shown in Figure 4.5. For this case, the probability of failure is given by

$$p_f = \int_{-\infty}^{\infty} F_R(x) f_Q(x) dx \qquad (4.6)$$

where $f_Q(x)$ is the probability density function of the load Q and $F_R(x)$ is the cumulative probability density function of the resistance ($F_R(x)$ is the probability that $R \leq x$). For the probability of failure of the R − Q system to be estimated, the probabilistic properties of R and Q need to be known. This will be considered in the following sections. However, before proceeding note that the failure probability may be calculated per annum, per lifetime or for any other time period. It follows that the probabilistic models selected for loads and resistances must relate to this time period. For example, the lifetime reliability for a communications tower requires that the probabilistic model of the maximum wind speed to be experienced over the lifetime of the structure is known. Further discussion of these matters is deferred to Chapter 6.

4.4.1 Modelling of resistances

The resistance, strength or capacity of structural, geotechnical, mechanical, hydraulic, electrical/electronic and other elements is influenced by:

1. material properties (e.g. steel yield stress);
2. geometric dimensions (e.g. thickness of containment vessel); or
3. values calculated from a predictive model.

The probabilistic models which describe the variability of the material and the geometric variables sometimes may be deduced directly from (the analysis of) appropriate statistical data. This is the case for probabilistic models for geometric and material variables. In other cases the model may need to be developed from probabilistic models for constitutive parts. For example, a probabilistic model for structural steel member resistance can be derived directly from probabilistic models for the steel material property and the geometric variables of the cross section. For the structural strength of a steel tension member the relationship (i.e. the model) connecting the variables is given by $N_t = A_g F_y$ where A_g is the cross-sectional area of the steel member (geometric variable) and F_y is the steel yield strength (material variable). Simple relationships from probability theory allow the mean, variance and higher moments, or the complete probability density function for N_t to be obtained from the probability distributions for A_g and F_y. Probabilistic models for some typical material and geometric variables are described in the following sections. The derivation of probabilistic models from models for constituent models will also be described.

4.4.2 Models for material and geometric variables

For structural steel elements the material properties of interest usually are the yield strength, the ultimate tensile strength, the modulus of elasticity, the shear modulus and Poisson's ratio. Statistical data for these variables has been obtained mainly from tests of billets produced at steel mills. A summary of statistical parameters for these variables is given in Table 4.8. The extreme value type I distribution, the lognormal distribution, and to a lesser extent, the truncated normal distributions may all be used to represent steel yield strength (Melchers, 1987) (see Figure 4.6). A summary of studies of the yield strength of reinforcing bars is reported by Mirza and MacGregor (1979a). There is relatively little available data for geometric or dimensional variables; however, some suggested statistical parameters are given in Table 4.9. For mechanical components, the size, location and clearances are important geometrical properties. Haugen (1980) has suggested that, in the absence of experimental data, (rather crude) probabilistic models can be derived from specification or tolerance limits.

Reliability of load-resistance sub-systems

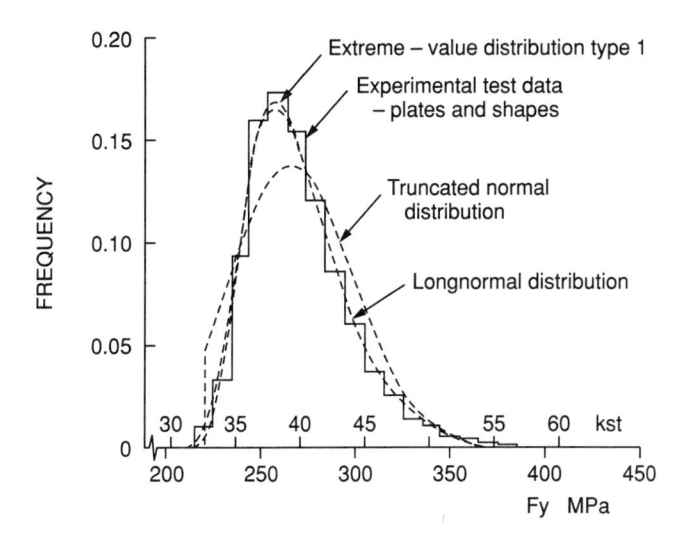

Figure 4.6 Typical histogram for yield strength with three fitted probability distributions.
Source: adapted from Alpsten (1972).

Table 4.8 Summary of statistical parameters for steel material properties

Material property	Mean/ specified	Coefficient of variation	Source
Modulus of elasticity	1.00	0.06	Galambos and Ravindra (1978)
Shear modulus	1.00	0.06	Galambos and Ravindra (1978)
Poisson's ratio	1.00	0.03	Galambos and Ravindra (1978)
Yield strength	1.05	0.10	Galambos and Ravindra (1978)
Tensile strength	1.10	0.11	AISI (1978)

Source: adapted from Ellingwood et al. (1980).

Table 4.9 Summary of statistical parameters for dimensional properties

Dimensional property	Mean/ specified	Coefficient of variation	Source
Flange thickness	0.97	0.028	O'Meara (1978)
Web thickness	1.00	0.032	O'Meara (1978)
Plate thickness – 6.4 mm	1.00	0.036	Baker (1970)
– 50 mm	1.00	0.0007	Baker (1970)
I-sections			
– section area	1.00	0.006	Pham and Bridge (1985)
– elastic modulus	0.97	0.05	Pham and Bridge (1985)
Initial deformation of plating	1.0	0.048	White et al. (1995)
Area of reinforcing steel	0.99	0.024	Mirza and McGregor (1979a)[a]

Note: [a] distribution truncated at 0.94.

Table 4.10 Average rates of corrosion of steel in sea water

Exposure level	Corrosion rate [milli-inches per year]
Deeply immersed	2–4
Ship's hull, immersed	5–7
Ship's hull, immersed in tropical or contaminated water	10–14
Ship's hull splash zone, highly oxygenated	up to 15

Source: adapted from Stiansen *et al.* (1980).

For some engineering systems it may be necessary to consider corrosion or wastage effects. For example, exposure of steel elements (as used in shipping, offshore platforms or industrial plants) to sea water or industrial chemicals may cause corrosion. The corrosion of steel elements is a complex phenomena; however, a simplistic and mathematically convenient approach is to assume that corrosion reduces the thickness of the exposed steel element by a similar amount each year. Stiansen *et al.* (1980) have summarized data relating to the average rates of corrosion of steel plates, see Table 4.10. However, this does not indicate the uncertainty in corrosion rates, which can be considerable (Melchers, 1995).

Metal fatigue accounts for more than 80% of all service failures in mechanical and structural elements. Fatigue requires crack initiation (e.g. from initial fabrication defects) and then crack propagation under environmental or cyclic loading conditions, sudden fracture then occurs. Probabilistic approaches to modelling fatigue and fracture have generated a large literature. Overviews have been given by Wirsching (1995), Harris (1995) and others.

A large amount of data has been collected for concrete compressive strength and some of this has been useful in developing probabilistic models. Some suggested statistical parameters for in-situ concrete compressive strength are available (Melchers, 1987). The influence of concrete quality and workmanship on probabilistic models of concrete compressive strength has been reported also (Stewart, 1995). Measurements of dimensional accuracy in reinforced concrete elements have been the principal means of obtaining data on the variation of the cross-sections of formed members and in the location of reinforcing bars. Summaries of statistical parameters for reinforced concrete beams, slabs and columns have been provided by Mirza and MacGregor (1979b) and Ellingwood *et al.* (1980). However, Reid (1987) has suggested that some of these suggested variances (e.g. of slab depths) may be overestimated because of deficiencies in the statistical methods used to determine variability of element dimensions.

The resistance of geotechnical systems may be measured in terms of stability of a slope in natural or man-made earth, resistance to sliding of dams, or resistance to other failure mechanisms. For soil, resistance is dependent on material and geometric variables such as sliding or slip area, angle of soil friction, cohesion, pore water pressure, and unit weight of soil. Uncertainties in these variables may be due to spatially varying soil properties and insufficient information or inadequate knowledge of the soil profile. The largest uncertainties are generally associated with the estimation of soil properties. Typically, these properties (e.g. angle of soil friction, cohesion) are obtained from laboratory tests on samples taken from the site. However, it is well known that laboratory measurements do not necessarily represent in-situ soil properties. Correction factors therefore are applied to the laboratory measured values: these factors being treated as random variables (Yucemen and Al-Homoud, 1990).

Probabilistic models that describe the physical properties of other man-made and natural materials such as metal alloys, fibre composites, ceramics, timber and glass are available from a large number of sources. For example, the properties of many metal alloys used in the aerospace industry are obtained from *Military Standardization Handbook: Metallic Materials and Elements for Aerospace Vehicles* (USAF, 1992). For each metal alloy, the handbook lists two property values corresponding to the 90% and 95% percentiles respectively. Standard statistical techniques could be used to infer statistical parameters such as the mean and variance, provided reasonable assumptions are made about the probability distribution which might apply.

For hydraulic systems a typical hydraulic resistance is the discharge capacity of a culvert. The resistance is dependent on a number of material and geometric variables; for example, upstream and downstream flow velocities, head loss due to friction in the pipe, pipe length, and pipe diameter. In some cases it may be difficult to identify the critical failure mechanism; for example, flow through a culvert may be modelled as either an open channel or a pressurized conduit, and the critical flow location may be upstream or downstream of the culvert (Yen, 1987). Statistical parameters may be obtained from site surveys, available literature or estimated from past experience.

The literature contains little information on the probability distributions of strength or capacity of electrical/electronic components. This is not unexpected since the reliability of components with small unit costs can be estimated directly from experimental failure data while the reliability of more expensive components, though produced in significant quantities, can be obtained from field data collection or incident data banks. However, probabilistic models of material and geometric properties would be needed if the reliability of a new component is to be predicted (Shooman, 1968).

It is important to note that statistical parameters for many geometric and material variables are obtained from measurements made from only limited numbers of samples. This suggests that it is therefore statistically unlikely that the limited samples will include the influence of gross errors.

4.4.3 Derived models for resistance

The resistance or capacity of some structural, geotechnical, hydraulic and other elements must be obtained by calculation from a known physical or other model. The outcome of such a model may be referred to as the 'nominal' resistance R_{nom}, even when the inputs are not necessarily nominal values. The reason for this is that the models often are those used for design purposes, and the results obtained from design models tend to be conservative. However, what is of interest in an engineering risk analysis is the actual resistance of the element or system (R). The actual resistance may be related to the nominal resistance by the expression

$$R = ME \times R_{nom} \qquad (4.7)$$

where ME is the 'modelling error' that corrects for the approximations made in calculating R_{nom}. For structural elements, the modelling error is obtained by conducting load tests on real members (e.g. destructive testing in a laboratory) to obtain statistical parameters of the ratio of actual or tested capacity to 'nominal' capacity. Typically, tests must be conducted on a variety of 'realistic' loading conditions and element sizes and these results are then aggregated so that statistical parameters for ME may be inferred. Figure 4.7 shows some results for steel beam buckling tests.

Variability due to uncertainties in the measured loads (e.g. errors in readings, accuracy of measuring devices), element geometric and material properties also must be considered (e.g. Ellingwood et al., 1980). The actual resistance may be expressed as

$$R = f(X_1, X_2, X_3, \ldots, X_N) \qquad (4.8)$$

where X_1 is the random variable for modelling error, and $X_2, X_3, \ldots, X_N$ describe the material and geometric properties needed to calculate ME and to calculate R_{nom}. To keep matters simple (and because of lack of better data) the latter random variables are often described only by their first and second moments (i.e. mean and variance). A probabilistic model of R can then be obtained by the Mean Value Method or Monte Carlo simulation. The Mean Value Method is most accurate when the function f() is linear. Monte Carlo computer simulation is generally used if modelling error, material or geometric probabilistic models are non-normal, if the function f() is complex or highly non-linear, or if the function f() is

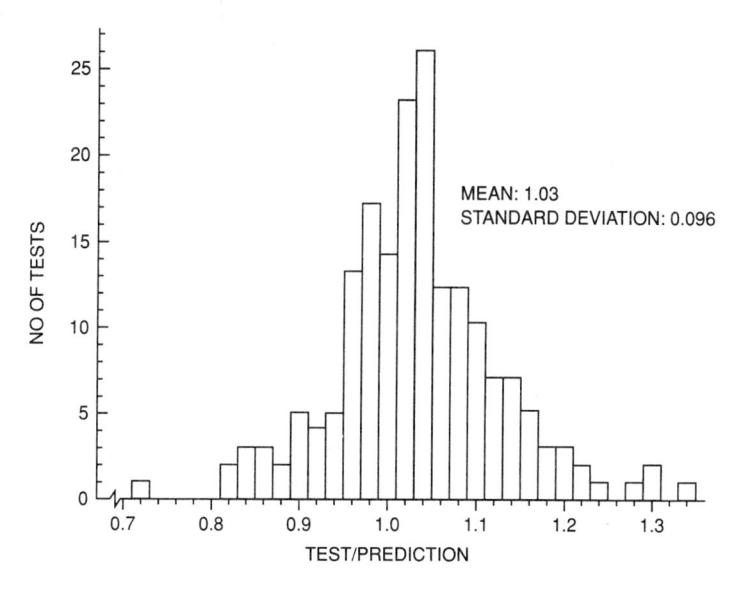

Figure 4.7 Histogram of 185 Beam Buckling Tests.
Source: Yura et al. (1978).

not available in a closed form solution. These methods are used extensively in system evaluation and are described in Chapter 6.

Probabilistic models of actual resistances are generally represented as the ratio of mean actual to nominal resistance ($\bar{R}/R_{nom}$) and the coefficient of variation ($V_R = \sigma_R \times \bar{R}/R_{nom}$). Some values for these statistical parameters are given in Table 4.11, for typical structural steel, reinforced concrete, geotechnical and hydraulic elements. In most cases, the distribution of resistance will be influenced by loading conditions, element size and mode of failure. For this reason, it is generally not possible to simply adopt generic distributions of resistance. For example, the parameters $\bar{R}/R_{nom}$ and V_R for reinforced concrete beams are significantly affected by concrete compressive strength, grade of reinforcing steel, element depth, proportion of reinforcing steel, and the predictive model used to calculate R_{nom}.

4.4.4 Modelling of loads

Loads or demands on a (sub-)system may be influenced by man-made causes or natural phenomena, and should be given in terms that enable direct comparison with the resistance or capacity of the system. Loads, particularly those resulting from natural phenomena tend to have a high degree of variability. In terms of recorded data, this is increased by measurement and recording variability. Many natural phenomena

Table 4.11 Summary of statistical parameters for structural resistance distributions

Element	Type of element	Details	$\bar{R}/R_{nom}$	V_R
Reinforced concrete:				
Flexure	Two-way slabs	depth = 125 mm	1.16	0.15
	Beams		1.01	0.12
Axial load/flexure	Short columns		1.05	0.15
	Long columns	Compression failure	1.10	0.17
		Tension failure	0.95	0.12
Shear	Beams	Min stirrups	1.00	0.19
Structural steel:				
Tension			1.10	0.11
Flexure	Plate girder		1.08	0.12
Hydraulic:				
Flow capacity	Culvert		1.00	0.08
Sliding resistance	Earth dam		1.00	0.10
Geotechnical:				
Slope stability	Embankment		–	0.194

Source: adapted from Ellingwood *et al.* (1980); Yen (1987); Bury and Kreuzer (1985); Yucemen and Al-Homoud (1990).

(wind, earthquakes) may be represented as stochastic processes. However, it is often more convenient for reliability studies to represent stochastic processes as time-invariant random variables, using maximum or peak values instead of all the observed values to represent the random variable. In the following the probabilistic representations of several important classes of loading will be described briefly; these are

1. earthquake loads,
2. flood loads,
3. wind loads,
4. live loads and
5. dead loads.

Two or more loads may act simultaneously on a system (e.g. dead and wind loads on a roof). This presents considerable complexity in appropriate representation (e.g. Melchers, 1987). However, it is a situation which analysts must understand and for this reason, a typical load combination technique will be described.

(a) Earthquake loading

Earthquake loading on structures or systems is clearly dynamic in nature. It will depend on a number of variables, such as:

1. peak ground acceleration,
2. duration of excitation,

3. mass of structure, or system,
4. ductility of materials, members and structural system,
5. spectral amplification,
6. support or (for buildings) the soil characteristics and
7. modelling error.

In most practical situations the uncertainties in the loading effects are 'overwhelmingly dominated' by the uncertainties in the peak ground acceleration. This applies to dynamic loading, but also for 'static equivalent' loading, a common artifice still common for the design of simpler systems, including relatively small buildings.

The variability and uncertainty in peak ground acceleration is relatively high due to the uncertainty associated with limited historical earthquake records and limited understanding of earthquake mechanics, wave travel path geology, and local soil conditions. For some seismic risk analyses (e.g. for nuclear power plants) the variation in peak ground acceleration is represented by a 'seismic hazard curve' (Ravindra, 1990; USNRC, 1989; Hwang et al., 1987). This curve is a cumulative distribution of the annual peak ground acceleration and represents the annual probability of exceeding a given value of peak ground acceleration. Because of the scarcity of data, particularly in relatively quiescent zones, probabilistic models and their statistical parameters are often elicited through expert judgement; measures of uncertainties in the parameter values and models can then be inferred (USNRC, 1989). Seismic hazard curves representing typical 15th, 50th and 85th percentiles are shown in Figure 4.8.

With reference to Figure 4.8 it is observed that there are substantial uncertainties associated with estimating peak ground accelerations. This is not unexpected since it has been shown in section 4.3.2 that expert opinions for component reliabilities can vary widely. This is the situation also for seismic characteristics.

Table 4.12 shows the range of peak ground accelerations obtained from 18 experts, with each expert supplied with the Modified Mercalli intensity (M_s) to be expected at each of seven hypothetical sites. In a review of this study, it was noted that some experts gave values below 0.05 g, yet it is known that the threshold of feeling anything in an earthquake is about 0.05 g; the low value 0.03 g is actually off the Modified Mercalli intensity scale. This led Krinitzsky (1993) to observe that 'Let's face it: experts can make very bad mistakes.' It would appear that this example is not an isolated one, and casts doubt over the use of multiple expert opinions to obtain probabilistic seismic hazard evaluations.

Energy absorption is often a more realistic measure of structural load since most structures exhibit inelastic behaviour. Assessing the energy absorbed by a structure is an extremely complex analytical procedure. Kuwamura and Galambos (1989) observe that uncertainties in the

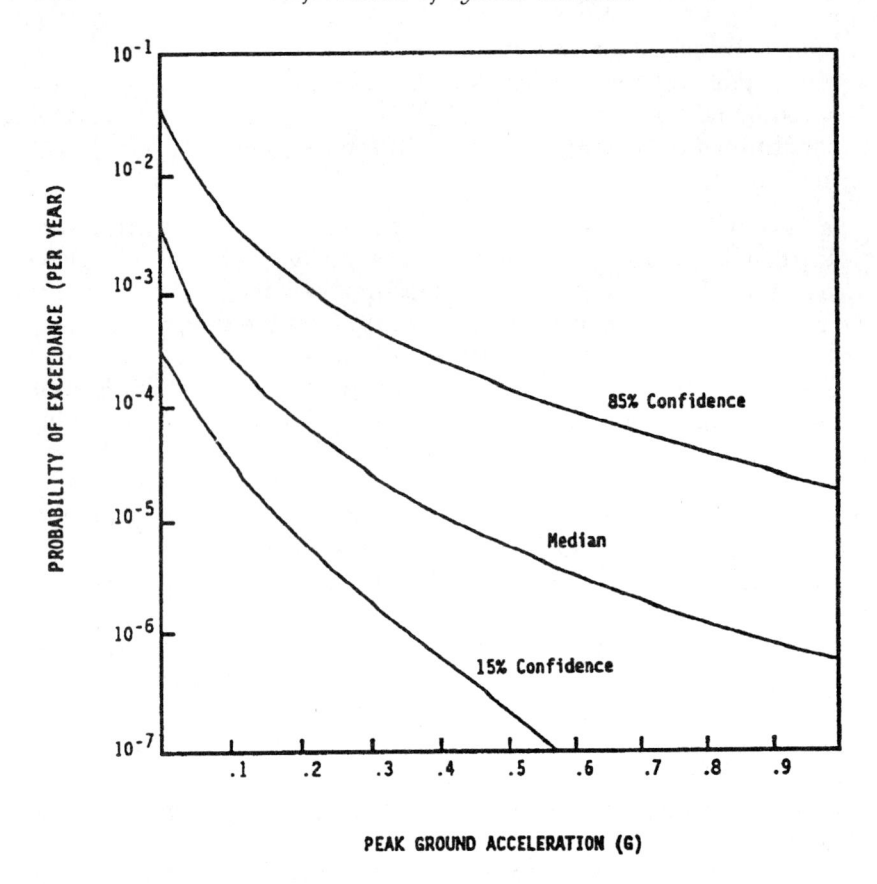

Figure 4.8 Seismic hazard curves for the Vermont Yankee Nuclear Power Plant.
Source: Ravindra et al. (1990).
Reprinted from M.K. Ravindra, M.P. Bohn, D.L. Moore and R.C. Murray (1990) Recent PRA applications, *Nuclear Engineering Design*, **123**, 155-66, with kind permission from Elsevier Science S.A., PO Box 564, 1001 Lausanne, Switzerland.

statistical properties of dynamic response characteristics are insignificant when compared to the maximum earthquake intensity uncertainties.

(b) Flood loading

Estimates of the probability of occurrence of flood loadings are required for the design of most hydrological systems such as dams, levee banks, culverts and bridge piers. The calculation of a flood load consists of the following computational steps:

1. estimation of rainfall intensity;

Table 4.12 Ranges in peak horizontal ground accelerations on soil by 18 experts

Location and site	Acceleration (g)	Velocity (cm/s)	Displacement (cm)	Duration (s)
San Andreas Fault (M_s=8.3)	0.35–3.0	46–550	40–300	20–90
5 km from San Andreas Fault (M_s=8.3)	0.35–3.0	46–550	20–300	20–90
50 km from San Andreas Fault (M_s=8.3)	0.18–0.4	20–100	10–40	20–50
150 km from New Madrid Source (M_s=7.5)	0.03–0.5	5–100	1–50	2–120
Reservoir Induced Earthquake (M_s=6.5)	0.35–2.0	40–300	20–190	10–30

Source: adapted from Krinitzsky (1993).

2. estimation of relative flood peak (i.e. discharge); and
3. conversion of discharge into a load or demand for direct comparison with the resistance or capacity of the system.

Some of the variables encountered when converting rainfall intensity to a yearly or 50-year flood peak are shown in Table 4.13. Each of the factors involved can be analysed in more detail. For example, the random variables associated with the estimation of rainfall intensity are

1. watershed (or catchment) size,
2. design period,
3. rainfall duration and
4. modelling errors.

Each of these should be represented by an appropriate probabilistic model derived from statistical properties obtained from past experience and observation. Similarly, there will be variability of the runoff coefficient is attributable variations of slope, vegetation, land use and antecedent conditions over the basin area. Unfortunately, there is some

Table 4.13 Variabilities in 50-year peak flood discharge prediction for a culvert design

Factor	Coefficient of variation
Rainfall intensity	0.149
Runoff coefficient	0.253
Basin area (1104 acres)	0.05
Modelling error	0.10
50 year peak flood discharge	0.314

Source: adapted from Yen (1987).

difficulty in obtaining good quality hydrological data. There are a number of reasons for this. A useful example is to note that typically only daily, and not hourly, or continuous readings of the flood levels or stages may recorded (Plate, 1984).

For some applications it is necessary to convert the actual flood discharge to a load, so that the load can be compared directly with the resistance of the hydrological system. This is done through load models, such as when the flood discharge into a dam is converted to a total hydrostatic head, which is then converted to a horizontal load in the plane of sliding at the base of the dam. These relationships are essentially empirical hence they have a degree of uncertainty, which, in some cases, is considerable. It has been suggested that the uncertainty in the load modelling (i.e. modelling error) is greater than the effect of uncertainty in rainfall intensity (Bury and Kreuzer, 1985).

There are some strong adherents to the use of the socalled Probable Maximum Precipitation (PMP) and Probable Maximum Flood (PMF) values for use in design and hence in risk analyses because these values are thought to represent extreme events (e.g. USBR, 1986). Unfortunately, frequencies of occurrence cannot be attached to these 'maxima' with any degree of certainty (Wellington, 1988). Risk analysts need to use such values with considerable care.

(c) Wind loads

The (equivalent) static wind pressure load acting on a structure typically is taken to be a function of

1. pressure coefficients (which depend on the size and geometry of a building and its orientation relative to predominent wind direction, etc.),
2. terrain exposure coefficient,
3. gust factor (e.g. turbulence), and
4. wind speed, referenced to a height of 10 m in open terrain.

Daily, weekly, yearly or 50-year (or higher) maximum wind pressure loads may be required for use in a risk analyses. (A 50-year wind has a probability of occurrence, on average of once every 50 years – the so-called 'return period' – see Chapter 6.)

Wind speeds are derived from meteorological data. This data is collected as either '3-second gust' speeds (which is the average wind speed over a gust, taken as of three seconds' duration), or, in the USA, as the 'fastest-mile' wind speed (which is the wind speed averaged over the passage of one mile). Clearly, these two measurements are not equivalent, but nonetheless, the respective data do provide similar estimates of mean hourly wind speeds. The extreme value type I distribution is recommended for daily and annual wind speeds (Ellingwood et al., 1980;

Simiu and Filliben, 1980; Pham *et al.*, 1983). The resulting daily maximum wind speeds may then be used to calculate 'arbitrary-point-in-time' loadings.

The maximum wind speed to be experienced during the lifetime of the structure is also of interest. This is referred to as the 'lifetime' maximum wind speed (with the lifetime typically taken as L = 50 years). The cumulative distribution function of lifetime maximum wind speeds ($F_{VL}(v)$) may be estimated from:

$$F_{VL}(v) = (F_V(v))^L \qquad (4.9)$$

where $F_V(v)$ is the cumulative distribution function for the annual maximum wind speeds and L is the lifetime in years. This assumes independence between annual maxima (e.g. Melchers, 1987).

Although there has been much investigation of wind speed and its intensity, relatively few statistical data have been collected for the variability of pressure coefficients, of the terrain exposure coefficients and of the gust factors. It has been suggested that some of these variables are correlated, but the extent of their correlation is uncertain. Ellingwood (1981) suggested that the coefficients of variation for pressure coefficients, terrain exposure coefficients and gust factors are in the range 0.10–0.15 and could be assumed to be normally distributed.

To determine wind force from wind velocity, it is noted that wind pressure (W) is directly proportional to the square of the value of the velocity (V). The appropriate probability distribution can be obtained only by Monte Carlo simulation because the square of an extreme value type I distribution has no closed form solution. It has been found by Ellingwood (1981) that the extreme value type I distribution produces the best fit for loads greater than the 90th percentile (see Table 4.14).

When an equivalent static load is not appropriate, such as for tall and slender structures and for sensitive mechanical equipment, it is necessary to consider dynamic effects. In this case, the wind load effects are described by the variables presented above and by additional variables that relate to structural properties and the fluid-interaction effect on the structure. These additional variables include spectral density of the wind velocity, stiffness and mass of the structure, structure damping

Table 4.14 Composite statistical parameters for wind loads

	$\overline{W}/W_{nom}$	V_W
50-yr maximum	0.78	0.37
Annual maximum	0.33	0.58
Daily maximum	0.01	7.00

Source: adapted from Ellingwood (1981).

coefficient, Reynolds number, and the Strouhal number. A description of the uncertainties associated with these variables is provided, for example, by Schueller et al. (1983).

It is important to note that the meteorological phenomena associated with cyclones and hurricanes is different from that associated with thunderstorms. Furthermore, cyclonic wind speeds are more difficult to predict because of the relatively limited amount of data available for these extreme events (Simpson and Riehl, 1981; Pham et al., 1983). The influence of long-term effects such as the El Nino phenomenon is also rather unclear.

(d) Live loads

Loads due to people, their possessions, furnishings, storage materials and other transient loads such as vehicles etc. are referred to as 'live' loads. The most common live load is floor loading acting on structural systems. This is now described in detail for illustration. Live loads may be divided into two categories:

1. sustained loads: long-term loads associated with normal use;
2. extraordinary loads: short-term transient loads caused by abnormal events.

The total live load is the sum of these two live load components. Each can be represented by a discrete stochastic process, as illustrated in Figure 4.9.

To develop probabilistic models for sustained and extraordinary live loads, live load surveys are required. However, owing to the high cost involved, only a few live load surveys have been conducted, even for important spaces such as offices (e.g. Culver, 1976; Choi, 1991). These surveys provided direct information on 'arbitrary-point-in-time' floor loadings.

The lifetime maximum sustained live load is influenced largely by changes in occupancy and/or room use during the life of the system. If it is assumed that the loadings for each occupant are independent and that changes in occupancy is represented by a Poisson counting process, then the cumulative distribution for the lifetime maximum sustained live load is

$$F_{Lm}(x) \approx \exp\{-v_0 t[1 - F_L(x)]\} \qquad (4.10)$$

where $F_L(x)$ is the cumulative distribution of the arbitrary-point-in-time sustained live load, v_o is the average rate of occupancy change, and t is the time period.

Data for 'extraordinary' live loads has been obtained through subjective estimates (e.g. a survey questionnaire) for the magnitude, the frequency of occurrence and the location of the extraordinary live loads;

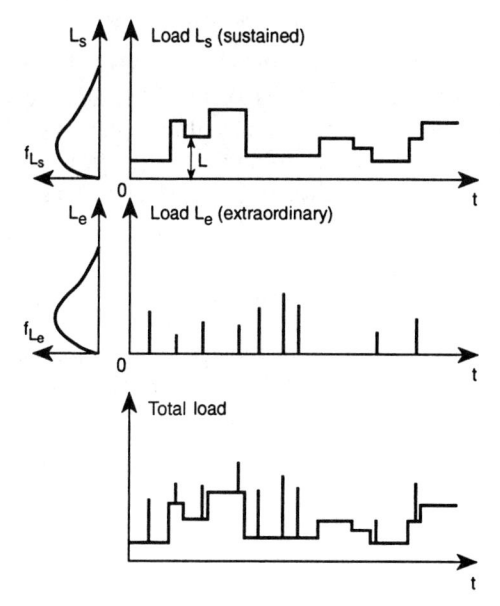

Figure 4.9 Time histories of typical live loads.
Source: Melchers (1987).

hence there is a considerable uncertainty about the data for these loadings. If the occurrence of individual extraordinary live loads is assumed to follow a Poisson counting process, the cumulative distribution of maximum extraordinary live loads may be approximated in a manner similar to that used in Eq. (4.10). Probabilistic models for office floor loads are shown in Table 4.15.

Probabilistic models for live load effects on other structural systems (for example, bridges, nuclear power plants, grandstands, and arenas) have been discussed in the literature (e.g. Nowak and Hong, 1991; Hwang et al., 1987; Saul and Tuan, 1986).

Table 4.15 Statistical parameters for some office floor loadings

Live load	Mean (kPa)	Coefficient of variation
Sustained live load:		
Arbitrary-point-in-time	0.53	0.70
Lifetime maximum (50 yrs)	1.21	0.27
Extraordinary live load:		
Single occurrence	0.39	1.03
Arbitrary-point-in-time (8 years)	1.20	0.31
Lifetime maximum (50 yrs)	1.79	0.23

Source: adapted from Chalk and Corotis (1980); Harris et al. (1981).

(e) Dead load

Dead load is generally taken as the self-weight of the system (or its subsystems or components). For structural elements, dead load may be modelled by the normal distribution, with a mean typically equal to the nominal design loads and a coefficient of variation of 0.06 to 0.15 (e.g. Ellingwood et al., 1980; Melchers, 1987). Most of this variability is due to uncertainties in non-structural loads.

(f) Other loads

Engineering systems may also be subject to other loads. The description of the relevant probabilistic models has been discussed in the literature; the following provides an overview:

1. containment pressure due to loss of coolant accident in a nuclear power plant (Hwang et al., 1987);
2. blast wave from an explosion in a chemical process plant (Lees, 1980);
3. thermal loads (Stiansen et al., 1980);
4. wave loading (Stiansen et al., 1980; White et al., 1995);
5. snow loads (Ellingwood and O'Rourke, 1983; Bennett, 1988);
6. aircraft, ship or vehicle impact loads (Lees, 1980);
7. pollutant loadings in an ecological system (Burges and Lettenmaier, 1975);
8. overloading of mechanical equipment (O'Connor, 1991); and even
9. meteoroid loading on lunar structures (Steinberg and Bulleit, 1994).

(g) Combination of load processes

When the system is subject to several time varying loads, acting at the same time, it is necessary to consider how these loads can be combined for use in a risk analysis. In general the analysis is complex. For most situations, however, simple, intuitive ideas can be adopted. For example, for structural systems, it is generally accepted that Turkstra's rule (Turkstra, 1970) is suitable for estimating the maximum load from a combination of loads. Turkstra's rule states that each combination of load effects consists of one load at its lifetime maximum and the others at their arbitrary-point-in-time (or instantaneous) or permanent values. Turkstra's rule is based on the not unrealistic assumption that it is highly unlikely that more than one maximum event will occur at any one point in time at the same time.

For other engineering systems also it is unlikely to be not unrealistic to assume that only one maximum event will occur at any one point in time. For example, for nuclear power plants it is recognized that it is overly conservative to combine the loads corresponding to the safe shutdown earthquake (SSE) load and the loads associated with a loss of coolant accident (LOCA) (Mattu, 1980). A further example is for dams, for which the risk of failure typically does not consider the simultaneous occurrence of flood and earthquake loads.

4.5 MODELLING OF CONSEQUENCES

The present chapter has concentrated on the probabilistic models which may be used to describe the behaviour and performance of individual components (or sub-systems or even complete systems). In some cases these models have to be derived from other models, as discussed in section 4.4. The probabilistic models are important ingredients in risk analyses since they are used to estimate the frequency of event failure and hence system failure (e.g. probability of structural collapse, aircraft crash or release of hazardous material). However, for most risk analyses, the system risk estimate also requires the estimation of the consequences of failure.

The consequences of a failure event typically are measured in terms of risks that directly affect people and their environment and may be influenced by one or more of the following variables:

1. nature of the hazard (e.g. toxic release, fire, explosion, structural collapse);
2. transport and dispersion of hazardous materials (e.g. toxic vapour cloud, radioactive materials);
3. meteorological conditions (e.g. may influence survival rates of shipping accidents);
4. human intake (or ingestion) rates of contaminated water, air and soil (e.g. LaGoy, 1987);
5. number of people exposed to hazard;
6. emergency response actions (e.g. effectiveness of emergency evacuations); and
7. health and environmental effects of hazard (e.g. early or latent cancers).

The consequences of most interest are those that tend to result from catastrophic or low probability/high-consequence events, such as sometimes associated with nuclear power plants. For these the consequence of main interest is the magnitude of containment release and the off-site societal consequences; for example, Figure 4.10 shows a typical relationship between iodine release and early fatalities. In Probabilistic Safety Assessments in the nuclear industry these stages in consequence analysis are referred to as Level 2 (containment release) and Level 3 (off-site societal consequences) (IAEA, 1992).

Catastrophic failures such as radioactive or vapour cloud releases from nuclear or chemical process plants are very few in number. It follows

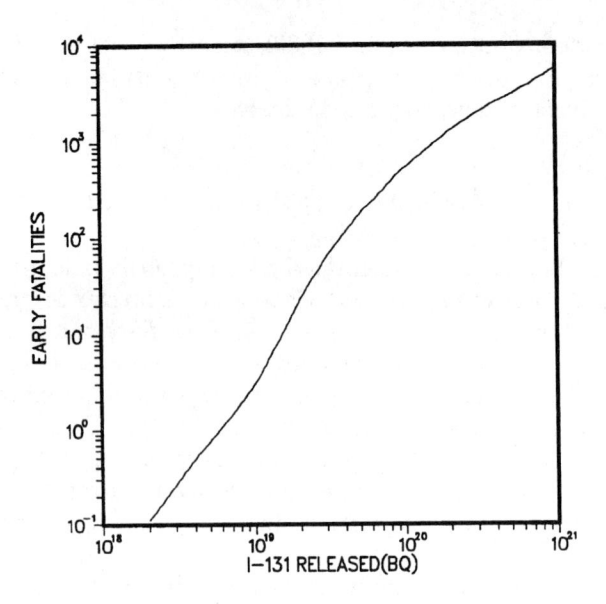

Figure 4.10 Relationship between I-131 release and mean early fatalities.
Source: USNRC (1989).

that the consequences of system failure for hazardous industries cannot be derived from direct observation alone. Given the absence of this raw data, predictive models need to be developed from experimental studies. For example, toxicology studies are based mainly on animal and laboratory experiments. The predictive models of the consequences of human exposure to toxins is essentially an extrapolation of these experimental studies and so are subject to considerable uncertainty. Additional uncertainties arise from the observation that only about 6% of all known potentially toxic chemicals (out of a total of 100 000) have been tested for carcinogenicity (mostly in animals); of these, only a small proportion have been found to be carcinogenic to humans. Since approximately 500 new chemicals are introduced each year, it may be some time (if ever) before the true level of toxicity of these new chemicals is known (Whittmore, 1983).

Not surprisingly, the consequences of failure are highly dependent on the engineering system and its mode of failure. For this reason, consequence analysis does not form part of the present book. Reference should be made to industry or discipline specific texts or literature. Details of some typical failure consequences and their quantitative representation are described elsewhere (e.g. Lees, 1980; Kaiser, 1982; USNRC, 1989).

4.6 SUMMARY

The measure of performance for components and items of equipment is reliability, and reliability data is often expressed as failure rates or mean time to failure. Failure may be due to internal (metal fatigue, wear, age etc.) or external (earthquake, flood, fire etc.) sources. Failure data obtained from laboratory testing, field data, incident data banks and expert opinions are used to estimate reliability data. This may be generic, industry or site specific and is generally contained or stored in reliability databases for convenient reference. Uncertainties in failure data result from, among other things, the definition of failure, non-standardized data reporting, who constitutes an 'expert' and limited sample sizes. Moreover, failure rates may vary from component to component. For these reasons it is preferable for reliability data to be presented in probabilistic terms as this helps reflect parameter uncertainties as well as the inherent variability of the component.

For some components or sub-systems, reliability data is not available through component testing. However, estimates of failure probability may be obtained by calculation using the 'load-resistance' or 'demand-capacity' approach, with element failure defined to occur when load exceeds resistance. This technique requires the probabilistic models of resistance, strength or capacity and load, demand or stress to be known. Probabilistically modelling these variables is complicated by the knowledge that some variables are stochastic processes; this includes natural phenomena such as wind, earthquakes and flooding. It is often more convenient to model these as stationary processes where the variable of interest may be the lifetime peak force. Appropriate load combination methods are needed when more than one load acts on a system.

REFERENCES

AISI (1978) Proposed criteria for load and resistance factor design of steel building structures, American Iron and Steel Institute, *AISI Bulletin*, **27**.

Alpsten, G.A. (1972) Variations in mechanical and cross-sectional properties of steel, tall building criteria and loading, Vol. Ib, *Proceedings of the International Conference on Planning and Design of Tall Buildings*, Lehigh University, Pennsylvania.

Anderson, T. and Misund, A. (1983) Pipeline reliability: an investigation of pipeline failure characteristics and analysis of pipeline failure rates for submarine and cross-country pipelines, *Journal of Petroleum Technology*, **35**(4), 709–17.

Ang, A.H.S and Tang, W.H. (1975) *Probability Concepts in Engineering Planning and Design, Volume 1 – Basic Principles*, Wiley, New York.

Apostolakis, G. (1982) Data analysis in risk assessments, *Nuclear Engineering and Design*, **71**, 375–81.

Apostolakis, G. (1986) On the use of judgement in probabilistic risk analysis, *Nuclear Engineering and Design*, **93**, 161–6.

Apostolakis, G., Kaplan, S., Garrick, B.J. and Duphily, R.J. (1980) Data specialization for plant specific risk studies, *Nuclear Engineering and Design*, **56**, 321–9.

Baker, M.J. (1970) *Variations in the Strengths of Structural Materials and Their Effect on Structural Safety*, Imperial College, London.

Balestreri, S. (1987) The component event data bank and its uses. In A. Amendola and A.Z. Keller (eds), *Reliability Data Bases: Proceedings of the ISPRA Course held at the Joint Research Centre, Ispra, Italy*, D. Rdeidel Publishing Co., Holland, 287–316.

Bello, G.C. (1987) Data validation procedures. In A. Amendola and A.Z. Keller (eds), *Reliability Data Bases: Proceedings of the ISPRA Course held at the Joint Research Centre, Ispra, Italy*, D. Rdeidel Publishing Co., Holland, 125–32.

Bennett, R.M. (1988) Snow load factors for LRFD, *Journal of Structural Engineering*, ASCE, **104**(10), 2371–83.

Burges, S.J. and Lettenmaier, D.P. (1975) Probabilistic methods in stream quality management, *Water Resources Bulletin*, **11**(1), 115–30.

Bury, K.V. and Kreuzer, H. (1985) Assessing the failure probability of gravity dams, *Water Power and Dam Construction*, **37**(11), 46–50.

Carter, A.D.S., Martin, P. and Kinkead, A.N. (1984) Design for reliability. In *Mechanical Reliability in the Process Industries*, Institution of Mechanical Engineers, London, 1–10.

CCPS (1989) *Guidelines for Process Equipment Reliability Data*, Center for Chemical Process Safety, American Institute of Chemical Engineers, New York.

Chalk, P.L. and Corotis, R.B. (1980) Probability models for design live loads, *Journal of the Structural Division*, ASCE, **106**(ST10), 2017–33.

Choi, E.C.C. (1991) Extraordinary live load in office buildings, *Journal of Structural Engineering*, ASCE, **117**(11), 3216–27.

Colombari, A.G. (1987) Data treatment in reliability data banks. In A. Amendola and A.Z. Keller (eds), *Reliability Data Bases: Proceedings of the ISPRA Course held at the Joint Research Centre, Ispra, Italy*, D. Rdeidel Publishing Co., Holland, 43–54.

Culver, C.G. (1976) *Survey Results for Fire Loads and Live Loads in Office Buildings*, National Bureau of Standards, NBS Building Science Series 85, Washington, DC.

Daniels, B.K. (1982) Data banks for events, incidents, and reliability. In A.E. Green (ed.), *High Risk Safety Technology*, Wiley, Chichester, 259–91.

Davidson, J. (1988) *The Reliability of Mechanical Systems*, Mechanical Engineering Publications Ltd, London.

Dhillon, B.S. (1988) *Mechanical Reliability: Theory, Models and Applications*, American Institute of Aeronautics and Astronautics Inc., Washington, DC.

DOD (1986) *Reliability Prediction of Electronic Equipment*, MIL-HDBK-217E, Department of Defense, Washington, DC.

Dowrick, D.J. (1987) *Earthquake Resistant Design* (2nd edn), Wiley-Interscience, Chicester.

Duphily, R.J. and Long, R.L. (1977) *Enhancement of Electric Power Plant Reliability Data Systems*, Fourth Annual Reliability Engineering Conference for the Electric Power Industry .

Ellingwood, B. (1981) Wind and snow load statistics for probabilistic design, *Journal of the Structural Division*, ASCE, **107**(ST7), 1345–50.

Ellingwood, B. and O'Rourke, M. (1983) Probabilistic models of snow loads on structures, *Structural Safety*, **2**, 291–9.

Ellingwood, B., Galambos, T.V., MacGregor, J.G. and Cornell, C.A. (1980) *Development of a Probability Based Load Criterion for American National Standard A58*, National Bureau of Standards Special Publication 577, US Government Printing Office, Washington DC.

References

Galambos, T.V. and Ravindra, M.K. (1978) Properties of steel for use in LRFD, *Journal of the Structural Division*, ASCE, **104**(ST9), 1459–68.

Garnier, N. (1987) Electronics data acquisition. In A. Amendola and A.Z. Keller (eds), *Reliability Data Bases: Proceedings of the ISPRA Course held at the Joint Research Centre, Ispra, Italy*, D. Rdeidel Publishing Co., Holland, 267–85.

Green, A.E. (1983) *Safety Systems Reliability*, Wiley, Chichester.

Harris, D.O. (1995) Probabilistic fracture mechanics. In C. Sundararajan (ed.), *Probabilistic Structural Mechanics Handbook: Theory and Industrial Applications*, Chapman & Hall, New York, 106–45.

Harris, M.E., Corotis, R.B. and Bova, C.J. (1981) Area dependent processes for structural live loads, *Journal of the Structural Division*, ASCE, **107**(ST5), 857–72.

Haugen, E.B. (1980) *Probabilistic Mechanical Design*, John Wiley & Sons, New York.

Hecht, H. and Hecht, M. (1986) Software reliability in the system context, *IEEE Transactions on Software Engineering*, **SE-12**(1), 51–8.

Henley, E.J. and Kumamoto, H. (1981) *Reliability Engineering and Risk Assessment*, Prentice Hall, New Jersey.

Hwang, H., Ellingwood, B. and Shinozuka, M. (1987) Probability-based design criteria for nuclear plant structures, *Journal of Structural Engineering*, ASCE, **113**(5), 925–42.

IAEA (1988) *Component Reliability Data for Use in Probabilistic Safety Assessment*, IAEA TECDOC 487, International Atomic Energy Agency, Vienna.

IAEA (1992) *Procedures for Conducting Probabilistic Safety Assessment of Nuclear Power Plants (Levels 1,2,3) – A Safety Practice*, Safety Series No. 50-P-4, International Atomic Energy Agency, Vienna.

IEEE (1984) *IEEE Guide to the Collection and Presentation of Electrical, Electronic, Sensing Component, and Mechanical Equipment Reliability Data for Nuclear-Power Generating Stations*, IEEE Std. 500-1984, Institute of Electrical and Electronic Engineers, New York.

Johnson, D.W. and Welker, J.R. (1981) *Development of an Improved LNG Plant Failure Rate Data Base*, Gas Research Institute, Chicago, Illinois.

Kaiser, G.D. (1982) Consequence assessment. In A.E. Green (ed.), *High Risk Safety Technology*, Wiley, Chichester, 93–100.

Kannegieter, T. (1993) In most cases it is a fallacy that equipment fails due to component wear, *Civil Engineers Australia*, July, 54.

Kaplan, S. (1992) 'Expert information' versus 'expert opinions': another approach to the problem of eliciting/combining/using expert knowledge in PRA, *Reliability Engineering and System Safety*, **35**, 61–72.

Klaassen, K.B. and Peppen, J.C.L. (1989) *System Reliability: Concepts and Applications*, Edward Arnold, London.

Krinitzsky, E.L. (1993) Earthquake probability in engineering – Part 1: the use and misuse of expert opinion, *Engineering Geology*, **33**, 257–88.

Kuwamura, H. and Galambos, T.V. (1989) Earthquake load for structural reliability, *Journal of Structural Engineering*, ASCE, **115**(6), 1446–62.

LaGoy, P.K. (1987) Estimated soil ingestion rates for use in risk assessment, *Risk Analysis*, **7**(3), 355–9.

Lees, F.P. (1980) *Loss Prevention in the Process Industries*, Vols 1 and 2, Butterworths, London.

Martz, H.F. (1984) On broadening failure rate distributions in PRA uncertainty analyses, *Risk Analysis*, **4**(1), 15–23.

Mattu, R.K. (1980) *Methodology for Combining Dynamic Responses*, Report No. NUREG-0484, Rev. 1, US Nuclear Regulatory Commission, Washington, DC.

McFadden, R.H. (1990) Developing a database for a reliability, availability, and maintainability improvement program for an industrial plant or commercial building, *IEEE Transactions on Industry Applications*, **26**(4), 735–40.

Melchers, R.E. (1987) *Structural Reliability: Analysis and Prediction*, Ellis Horwood, Chichester, England.

Melchers, R.E. (1995) Probabilistic modelling of marine corrosion of steel specimens, *Proc. 5th International Offshore and Polar Engineering Conference, the Hague, the Netherlands*, ISOPE, 205–10.

Mirza, S.A. and MacGregor, J.G. (1979a) Variability of mechanical properties of reinforcing bars, *Journal of the Structural Division*, ASCE, **105**(ST5), 921–37.

Mirza, S.A. and MacGregor, J.G. (1979b) Variations in dimensions of reinforced concrete members, *Journal of the Structural Division*, ASCE, **105**(ST4), 751–66.

Mosleh, A. (1986) Hidden sources of uncertainty: judgement in the collection and analysis of data, *Nuclear Engineering and Design*, **93**, 187–98.

Nowak, A.S. and Hong, Y-K. (1991) Bridge live-load models, *Journal of Structural Engineering*, ASCE, **117**(9), 2757–67.

O'Connor, P.D.T. (1991) *Practical Reliability Engineering*, John Wiley & Sons, Chichester, UK.

O'Meara, G. (1978) *Quality Control of Structural Steel*, Project Report, Dept. of Civil and Aeronautical Engineering, Royal Melbourne Institute of Technology, Melbourne, Australia.

OREDA (1984) *Offshore Reliability Data Handbook*, OREDA, Veritec/PennWell Books, Hovik, Norway.

Paté-Cornell, M.E. (1986) Probability and uncertainty in nuclear safety decisions, *Nuclear Engineering and Design*, **93**, 319–27.

Pham, L. and Bridge, R.Q. (1985) Safety indices for steel beams and columns designed to AS 1250-1981, *Civil Engineering Transactions*, Institution of Engineers Australia, **CE27**(1), 105–10.

Pham, L., Holmes, J.D. and Leicester, R.H. (1983) Safety indices for wind loading in Australia, *Journal of Wind Engineering and Industrial Aerodynamics*, **14**, 3–14.

Pirovano, G. and Turconi, G. (1988) Telecommunications reliability databank: from components to systems, *IEEE Journal on Selected Areas in Communications*, **6**(8), 1364–70.

Plate, E.J. (1984) Reliability analysis of dam safety. In W.H.C. Maxwell and L.R. Beard (eds), *Frontiers in Hydrology*, Water Resources Publications, Littleton, Colorado, 288–304.

Ravindra, M.K. (1990) System reliability considerations in probabilistic risk assessment of nuclear power plants, *Structural Safety*, **7**, 269–80.

Ravindra, M.K., Bohn, M.P., Moore, D.L. and Murray, R.C. (1990) Recent PRA applications, *Nuclear Engineering and Design*, **123**, 155–66.

Reid, S.G. (1987) Variations from the design thickness of reinforced floor slabs, *First National Structural Engineering Conference*, Institution of Engineers Australia, Melbourne, 412–17.

RSS (1975) *An Assessment of Accident Risks in U.S. Nuclear Power Plants*, United States Nuclear Regulatory Commission, WASH-1400, NUREG-75/014, Appendix III.

Saul, W.E. and Tuan, C.Y. (1986) Review of live loads due to human movements, *Journal of Structural Engineering*, ASCE, **112**(5), 995–1004.

Schueller, G.I., Hirtz, H. and Booz, G. (1983) The effect of uncertainties in wind load estimation on reliability assessments, *Journal of Wind Engineering and Industrial Aerodynamics*, **14**, 15–26.

Shooman, M.L. (1968) *Probabilistic Reliability – An Engineering Approach*, McGraw-Hill Book Co.

Simiu, E. and Filliben, J.J. (1980) Weibull distributions and extreme wind speeds, *Journal of the Structural Division*, ASCE, **106**(ST12), 491–501.

Simpson, R.H. and Riehl, H. (1981) *The Hurricane and its Impact*, Louisiania State University Press.

Steinberg, E.P. and Bulleit, W. (1994) Reliability analyses of meteorite loading on lunar structures, *Structural Safety*, **54**, 51–66.

Stewart, M.G. (1995) Workmanship and its influence on probabilistic models of concrete compressive strength, *ACI Materials Journal*, American Concrete Institute, **92**(4), 361–72.

Stiansen, S.G., Mansour, A., Jan, H.W. and Thayamballi, A. (1980) Reliability methods in ship structures, *Transactions of the Royal Institution of Naval Architects*, **122**, July, 381–97.

Tomic, B. and Lederman, L. (1989) Data selection for probabilistic safety assessment. In V. Colombari (ed.), *Reliability Data Collection and Use in Risk and Availability Assessment: Proceedings of the Sixth EuroData Conference*, Springer-Verlag, Berlin, 80–9.

Turkstra, C.J. (1970) *Theory of Structural Design Decisions Study No. 2*, Solid Mechanics Division, University of Waterloo, Waterloo, Ontario.

USAF (1992) *Military Standardization Handbook: Metallic Materials and Elements for Aerospace Vehicles*, MIL-HDBK-5F, US Air Force, Wright Patterson Air Force Base, Ohio.

USBR (1986) *Guidelines to Decision Analysis*, ACER Technical Memorandum No. 7, United States Bureau of Reclamation, Denver, Colorado.

USNRC (1989) *Severe Accident Risks: An Assessment for Five Nuclear Power Plants*, NUREG-1150, US Nuclear Regulatory Commission, Washington, DC.

Verwoerd, M. (1991) Reliability data for use in offshore formal safety assessment studies. In R.F. Cox and M.H. Walter (eds), *Offshore Safety and Reliability*, Elsevier Applied Science, London, 90–104.

Vesely, W.E. and Rasmuson, D.M. (1984) Uncertainties in nuclear probabilistic risk analysis, *Risk Analysis*, **4**(4), 313–22.

Wellington, N.B. (1988) Dam safety and risk assessment procedures for hydrologic adequacy reviews, *Civil Engineering Transactions*, Institution of Engineers Australia, **CE30**(5), 318–26.

White, G.J., Ayyub, B.M., Nikolaidis, E. and Hughes, O.F. (1995) Applications in ship structures. In C. Sundararajan (ed.), *Probabilistic Structural Mechanics Handbook: Theory and Industrial Applications*, Chapman & Hall, New York, 575–607.

Whittmore, A.S. (1983) Facts and values in risk analysis for environmental toxicants, *Risk Analysis*, **3**(1), 23–33.

Wirsching, P.H. (1995) Probabilistic fatigue analysis. In C. Sundararajan (ed.), *Probabilistic Structural Mechanics Handbook: Theory and Industrial Applications*, Chapman & Hall, New York, 146–65.

Yen, B.C. (1987) Engineering approaches to risk and reliability analysis. In Y.Y. Haimes and E.Z. Stakhiv (eds), *Risk Analysis and Management of Natural and Man-Made Hazards*, ASCE, New York, 22–49.

Yucemen, M.S. and Al-Homoud, A.S. (1990) Probabilistic three-dimensional stability analysis of slopes, *Structural Safety*, **9**(1), 1–20.

Yura, J.A., Galambos, T.V. and Ravindra, M.K. (1978) The bending resistance of steel beams, *Journal of the Structural Division*, ASCE, **104**(ST9), 1355–70.

Human error and human reliability data

5.1 INTRODUCTION

Engineering systems involve individuals engaged on various individual tasks, using documents prepared by others and engaged in group activities of various types. Taken as a whole the process is a complex one. There is interaction between people and between people and their environment. People involved in this process may include planners, managers, engineers, draftsmen, construction workers, operators, inspectors, regulatory inspectors and others. Human error can occur from one or more of these participants or in the interaction between them. It has been suggested that all engineering systems are socio-technical systems and that it is not surprising that humans have been found to be the 'weakest link' in these systems (Turner, 1978). For example, human error causes 20–90% of all major system failures or accidents, see Table 5.1. As noted in Chapter 2 human error can occur in the following tasks:

1. administration/management,
2. design,
3. assembly/installation/construction,
4. operation/utilization,
5. inspection, and
6. maintenance.

The incidence of human error is not insignificant; for example, in one year there were 1345 reported 'abnormal occurrences' in US Commercial Boiling Water Reactor (nuclear) plants (Ujita, 1985). Moreover, it has been estimated that the proportion of failures in hazardous systems attributable to human error has increased from 20% in the 1960s to more than 80% in the 1990s (Hollnagel, 1993). For these reasons, the inclusion of human error into the computation of system reliability is necessary if 'realistic' estimates of system risk are to be calculated.

This chapter is concerned with human error and its quantification. As a starting point, a working definition of human error is provided. This

Introduction

Table 5.1 Proportion of system failures due to human error

System	Percentage of failures/accidents	Source
Aircraft	60–70%	Christensen and Howard (1981)
Air traffic control	90%	Kinney et al. (1977)
Buildings and bridges	75%	Matousek and Schneider (1977)
Dams	75%	Loss and Kennett (1987)
Missiles	20–53%	Christensen and Howard (1981)
Off-shore platforms	80%	Bea (1989)
Power plants: fossil-fuel	20%	Finnegan et al. (1980)
nuclear	46%	Scott and Gallaher (1979)
Shipping	80%	Gardenier (1981)

is followed by descriptions of several classifications of human error, their causes and their control. The distinction between errors and deliberate violations is explained also. This understanding, in conjunction with a Hazard Scenario Analysis discussed in Chapter 3, may then be used to provide an indication of potential errors that can occur in a specific system.

A Human Reliability Analysis (HRA) may be used to include the influence of human error (by operators, designers etc.) in the computation of system risk (e.g. Dougherty and Fragola, 1988; Kirwan, 1994). The need for HRAs developed initially from risk analyses of nuclear power plants (in the 1960s) when it was realized that the majority of accidents were due to human intervention (e.g. operator errors) and not simply due to equipment malfunctions as assumed in traditional risk analyses. Therefore, the aims of a HRA are to (1) provide an estimate of the human error contribution to system reliability and (2) incorporate this information into an overall risk analysis (one which includes the effect of equipment failure and other failure possibilities). Of course, human influences can be included directly in conventional events trees and fault trees composed of events involving mechanical and other components. This may be more appropriate for some analyses than performing a separate HRA.

A HRA uses event tree logic (see Chapter 3) to represent individual events (tasks) such as steps or operations required for the successful completion of the system task. Typical events may include opening a valve, a calculation, starting a pump or supervising an operator. Human reliability data for error rates, error magnitudes and/or error recovery will be required for these and other human actions for input into a Human Reliability Analysis. Several sources of human reliability data will be described, including human reliability databases and the use of expert opinions. Finally, the accuracy and validity of these sources of data are discussed.

5.2 HUMAN ERROR AND HUMAN BEHAVIOUR

Due to the nature of human behaviour and human error, there is no single classification scheme that can be applied to all tasks involving human errors, their causes or their control. For this reason, a taxonomy of errors usually is developed only for specific tasks. Nonetheless, several, often complementary, classifications that are relevant to engineering systems will be described below. More detailed descriptions of human error, their causes and their control are provided by Reason (1990, 1995); Rasmussen et al. (1987); and Senders and Moray (1991).

5.2.1 Definition of human error

What actually constitutes a 'human error' is somewhat subjective. However, a good working definition has been provided by Rigby (1970) who has defined human error as a human action that exceeds some limit of acceptability (or some allowable tolerance). For engineering professionals, this definition may be interpreted as an event or process that departs from commonly accepted competent professional practice (Melchers, 1984). Human error may be caused by 'error likely' conditions, such as poor motivation, time pressure, inexperience etc. (Rouse, 1985). In other words, human error is not simply due to wilful negligence, but is more likely to occur as a result of the unpredictable nature of human behaviour.

It is important to recognize that human errors are not necessarily detrimental to system safety; some errors actually increase system reliability. For example, Perrow (1984) cites the following case. An operator error in a nuclear power plant caused the reactor to automatically shut-down; technicians then discovered that 450 000 litres of water had leaked into the containment room. This leak was not discovered earlier because of equipment and sensor failure. Another example is that over-estimating stresses on a beam (e.g. caused by a calculation error) will lead to the selection of a larger than required beam.

5.2.2 Classification of human error

(a) Slips and mistakes

Human errors may be categorized broadly as either slips (or lapses) or mistakes (Norman, 1981). A slip is defined as a failure to achieve a set goal (e.g. a calculation error, forgetting to open a valve, or inadvertently closing the wrong valve) and a mistake as selecting an inappropriate goal (e.g. using an incorrect operating procedure, selecting an unsuitable design loading combination). In other words, a slip is an unconscious error and a mistake is an error due to a deliberate or conscious action.

(b) Physiological, psychological and philosophical factors

It is reasonable to assume that a correlation exists between human error and human behaviour. For this reason, Rasmussen (1979) has classified the behaviour of human operators for complex systems as

1. skill-based,
2. rule-based, or
3. knowledge-based.

These three categories provide an indication of the mental activities of the operator and apply to psychomotor and cognitive tasks.

Skill-based behaviour is essentially an 'automated' behaviour associated mainly with the execution of psychomotor or simple cognitive tasks; for example, operating a crane or simple arithmetic tasks. This behaviour generally does not involve any conscious effort. Rule-based behaviour requires a more conscious effort such as where memorized (or written) procedures must be followed (e.g. complex arithmetic tasks, use of design codes, operating procedures). This type of behaviour requires a longer response time and is more prone to error than is skill-based behaviour; errors may be due to lack of memory or lack of willingness to act on the part of the person concerned. Knowledge-based behaviour involves a more complex cognitive process and is associated with conscious solution of problems in unfamiliar situations (e.g. diagnosis of a problem during an emergency situation, the analysis of a novel structure or the evaluation of design changes). The higher degree of complexity involved on this type of activity would increase both the response time and the likelihood of error.

Reason (1990) has proposed incorporating Norman's two error types with Rasmussen's three levels of behaviour. The resulting Generic Error Modelling System (GEMS) has all errors defined in one of the following three basic error types (see Figure 5.1):

1. skill-based slips or lapses;
2. rule-based mistakes; or
3. knowledge-based mistakes.

In this scheme, skill-based slips tend to precede the occurrence of a problem (i.e. slip may cause a problem to develop). On the other hand, rule-based and knowledge-based mistakes are more likely to arise during subsequent attempts to find a solution (i.e. aware that a problem exists). Errors that result from skill-based slips and rule-based mistakes are generally predictable; for example, selecting an adjacent button or using other rules that could be satisfied (either partially or fully) by the situation under consideration. However, it is more difficult to predict

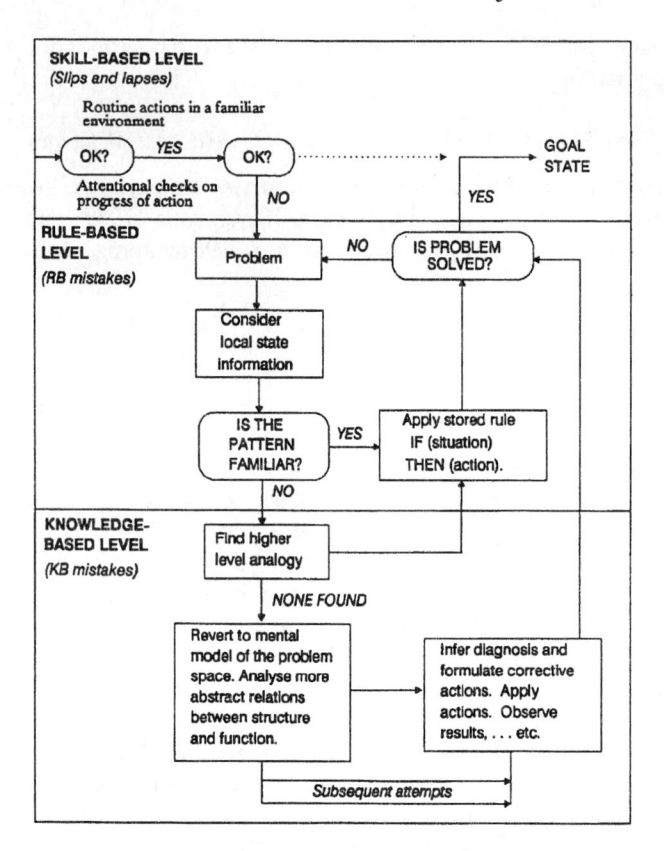

Figure 5.1 Outlining the dynamics of the Generic Error-Modeling Systems (GEMS).
Source: Reason (1990).

the outcome of knowledge-based mistakes. The outcome is likely to be influenced by the past experience of individuals and their knowledge of the system. It follows that it is reasonable to assume that knowledge-based mistakes can sometimes lead to 'unforeseen' consequences.

It is generally accepted that mistakes are harder to detect than slips (e.g. Reason, 1990); hence the prevention and control of these errors require differing error control strategies. For example, Wood (1984) has shown that nearly two-thirds of all operator errors which occurred in nuclear power plant were undetected in simulated plant failures. Approximately 50% of skill-based slips were detected by the operators themselves. However, none of the rule-based or knowledge-based mistakes were detected by the operators; these errors were only detected by other personnel (e.g. supervisors).

The three factors that contribute to human error in civil engineering tasks may be classified as:

1. physiological,
2. psychological, and
3. philosophical.

According to Ingles (1979) physiological factors tend to cause errors of sensing (e.g. failure to detect a signal) due to physical or environmental influences such as personal health, work overload (or underload), extreme temperatures, noise or other distractions, task complexity or 'user-unfriendly' man-machine interfaces. Errors of identifying and interpreting (e.g. incorrect choice of procedure) are generally caused by psychological factors such as lack of experience or knowledge, attitude and temperament. Furthermore, it is not uncommon to encounter tasks or problems that require a subjective judgement because the performance criteria are not clear or are contradictory. Typical examples include the precise delineation between two different soil types or the selection of an 'appropriate' safety factor. Errors in the execution of these tasks or problems are referred to by Ingles (1979) as philosophical errors.

(c) Organizational errors

Because of their size and complexity, many engineering systems tend to be operated and regulated by large organizations, with many employees. In this situation, it is highly likely that individuals seldom act in isolation, but tend to function within an environment that is controlled by one or more groups or other organizations. It is therefore not surprising that case studies of system failures indicate that most failures are influenced by organizational factors (e.g. Turner, 1978; Paté-Cornell, 1989; Reason, 1995). In a study of offshore platforms, Paté-Cornell (1989) categorized errors as either procedural or organizational. Procedural errors arise when an individual fails to perform a specific task in a procedure of operations that is assumed to be adequate. Therefore, the cause of any accident can be sourced directly to one or more identifiable initiating events (e.g. operator error). However, the initiating event itself could be due to organizational errors, such as those resulting from poor (or bad) organizational or management decisions; for example, excessive risk taking (perhaps due to financial constraints); communication problems between individuals, organizations and users; static organizational structure (may not be appropriate for new technology); reorganization and rationalization for economic reasons; or lack of incentive. A taxonomy of organizational errors is shown in Figure 5.2. These organizational errors may be closely associated with the philosophical errors described by Ingles (1979). Both these types of error occur in the design, construction and operation of offshore platforms (Paté-Cornell, 1989) and, it would be reasonable to assume, of other engineering systems (e.g. Turner, 1978).

An example of an organizational error in the management of administrative procedures for hazardous waste disposal is provided by

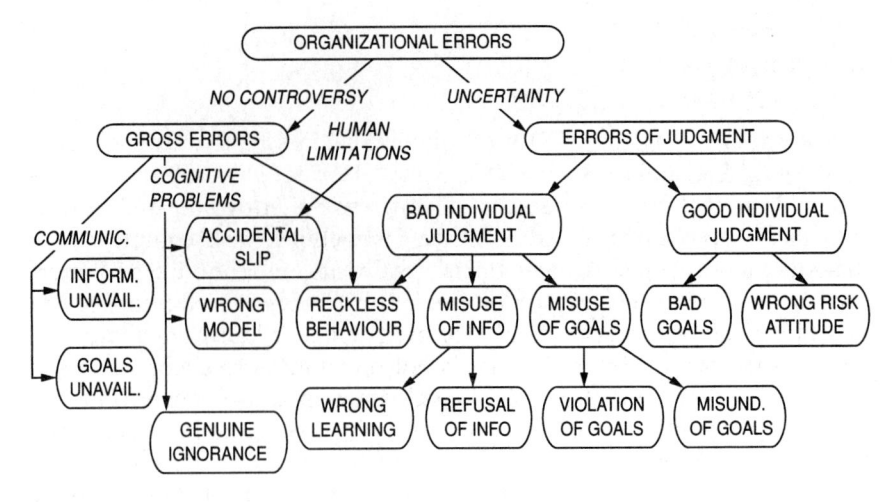

Figure 5.2 A taxonomy of organizational errors.
Source: Paté-Cornell (1989).

Sheppard (1992). Current regulations require that a six part carbonated form be used to record the processes involved in the carriage and disposal of hazardous waste. This form is to be filled in progressively and one carbon copy retained by each of the waste producer, carrier, and disposer. The disposal site copy is completed by the carrier and must be kept for perpetuity by the waste disposer. However, Sheppard (1992) observes that 'this (form) is the bottom most sheet and therefore the most illegible. Couple this with the fact that these things are often filled out in the rain, with a blunt pencil resting on the bonnet of a dirty lorry and it's no wonder that 90% of the pinks (forms) are unreadable!'. This is clearly an organizational error resulting in poor final documentation with severe potential safety related problems, particularly when existing disposal sites are considered for future development.

(d) Active and latent errors

Finally, it is useful to distinguish between errors that produce immediate consequences and errors that only become evident after a long period of time (i.e. errors are dormant and occur long before an accident sequence begins). These two error types are referred to by Reason (1990) as active errors and latent errors respectively. Active errors typically are committed by individuals who are in direct control of the system, such as pilots, ships officers, control room operators and air traffic controllers. However, latent errors are likely to be committed by those responsible for the physical and environmental characteristics of the system. These include high-level decision-makers, managers, engineers, construction workers and maintenance personnel. Evidently, many latent errors may be categorized as organizational errors. It is likely also that undetected latent errors contribute to the incidence of active errors because the appearance of latent errors often can place operators in an 'out-of-tolerance' system state (i.e. an unexpected emergency situation or one with which they are not familiar). Detailed analyses of recent accidents (e.g. Three Mile Island, Chernobyl, Bhopal and Challenger) have lead Reason (1990) to conclude that these system failures were primarily caused by latent errors in the systems. It appears that latent errors are a major contributor to system risk.

5.2.3 Performance Shaping Factors

Performance Shaping Factors (PSFs) are used to describe factors that influence human performance, either of the individual or of the organization. Miller and Swain (1987) have divided PSFs into two distinct categories: (1) external and (2) internal. External PSFs relate to the nature of the physical environment or task situation and are generally outside the control of the individual (e.g. characteristics of man-machine interface) while internal PSFs are those that involve the attributes, skills and abilities of the individual. It is observed that the physiological and psychological factors described by Ingles (1979) are directly analogous to external and internal PSFs respectively. A summary of some typical external and internal PSFs is given in Table 5.2.

In a study that assessed human errors in the operation of chemical process plants, it was found that approximately 87% of human errors were due to ergonomic design, managerial, and operational and environmental factors (see Table 5.3); the remaining errors were attributed to the character of the individual operator (Hayashi, 1985). These results support the view that human error is more likely to be caused by external PSFs. A further example of external PSFs is the control room environment during the accident at the Three Mile Island nuclear power plant:

Table 5.2 Examples of typical performance shaping factors

External PSFs	Internal PSFs
Inadequate work layout	Experience/skill level
Poor environment conditions	Stress level
Inadequate man–machine interface	Intelligence
Inadequate training	Motivation/attitude
Poorly defined job tasks	Emotional state
Poor supervision	Perceptual abilities
Task complexity	Physical condition
Time pressure	Social factors

Source: modified from Miller and Swain (1987).

Table 5.3 Human error-inducive situations

Human error-inducive situation	Frequency (%)
Ergonomic design	39.3
Operational and environmental	24.7
Managerial	23.3
Operator	12.7

Source: Hayashi, (1985).

110 alarms were sounding; key indicators were inaccessible; repair-order tags covered the warning lights of nearby controls; the data printout on the computer was running behind (eventually by an hour and a half); key indicators malfunctioned; the room was filling with experts; and several pieces of equipment were out of service or suddenly inoperative (Perrow, 1982).

It should not be unexpected that an environment such as this would induce errors.

5.2.4 Violations

A factor contributing to the structural failure of many buildings is the violation of codes, regulations, drawings, manuals and recommendations (Matousek and Schneider, 1977). Violations were also one of the causes of the accident at Chernobyl (Reason, 1990). It is reasonable to assume that violations also occur in the design, construction, and operation of other engineering systems. Clearly, it is important to distinguish between errors and violations. Unlike errors, which generally are caused by cognitive processes of the individual, violations can occur when behaviour is influenced by operating procedures, codes of practice and rules (Reason, 1990). A flowchart describing the classification of errors and violations is shown in Figure 5.3.

Violations may be defined as 'deliberate – but not necessarily reprehensible – violations from those practices deemed necessary (by designers, managers and regulatory agencies) to maintain the safe operation of a potentially hazardous system' (Reason, 1990). Two types of violations can be distinguished. Routine violations are those in which an action is based on least effort and/or the knowledge that the system rarely punishes violations (or rewards observances, for that matter). Fortunately, in many cases it is possible to identify activities which are susceptible to routine violations. Exceptional violations are not so predictable; they are influenced by unexpected local conditions or operating circumstances that make violations inevitable. For example, in an emergency situation, safety systems may be bypassed by the operators

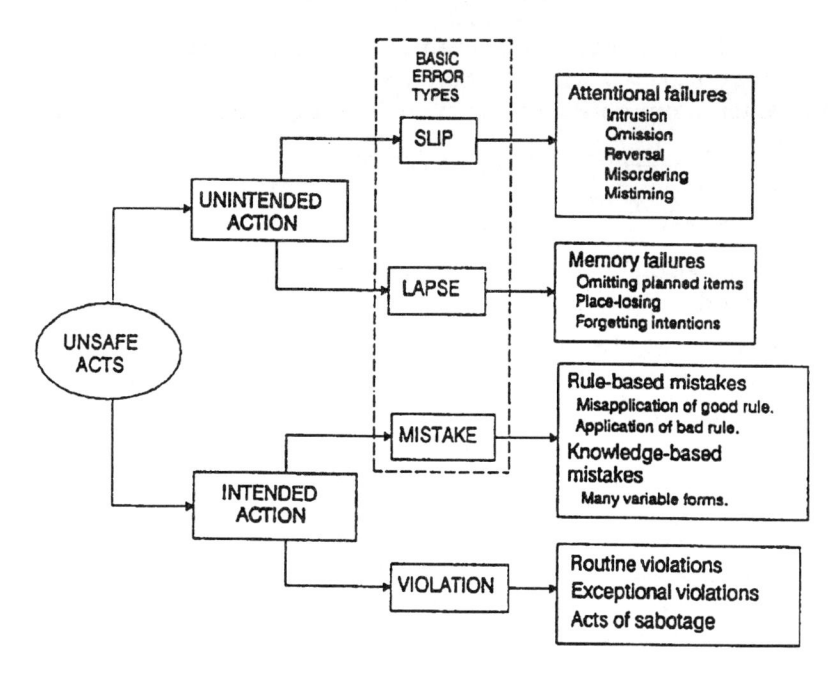

```
                    BASIC
                    ERROR
                    TYPES
                                        ┌─────────────────────┐
                   ┌───────┐            │ Attentional failures│
                ┌─▶│ SLIP  │───────────▶│ Intrusion           │
                │  └───────┘            │ Omission            │
   ┌──────────┐ │                       │ Reversal            │
   │UNINTENDED│─┤                       │ Misordering         │
   │  ACTION  │ │                       │ Mistiming           │
   └──────────┘ │                       └─────────────────────┘
         │      │                       ┌─────────────────────┐
         │      │  ┌───────┐            │ Memory failures     │
         │      └─▶│ LAPSE │───────────▶│ Omitting planned items│
         │         └───────┘            │ Place-losing        │
 ┌─────┐ │                              │ Forgetting intentions│
 │UNSAFE│┤                              └─────────────────────┘
 │ ACTS ││                              ┌─────────────────────┐
 └─────┘ │                              │ Rule-based mistakes │
         │         ┌────────┐           │ Misapplication of good rule.│
         │      ┌─▶│MISTAKE │──────────▶│ Application of bad rule. │
         │      │  └────────┘           │ Knowledge-based     │
         │      │                       │ mistakes            │
         │ ┌──────────┐                 │ Many variable forms.│
         └▶│ INTENDED │                 └─────────────────────┘
           │  ACTION  │                 ┌─────────────────────┐
           └──────────┘                 │ Routine violations  │
                  │    ┌─────────┐      │ Exceptional violations│
                  └───▶│VIOLATION│─────▶│ Acts of sabotage    │
                       └─────────┘      └─────────────────────┘
```

Figure 5.3 Classification of errors and violations.
Source: Reason (1990).

because of conflict between particular operating procedures – a matter of interest in the design of these systems.

In themselves, violations generally will not lead to system failure since most failures are due to multiple causes. For example, a study of accidents to railway shunters revealed that 91% of fatalities arise from being caught between vehicles while coupling or uncoupling or by being struck by walking on the line (Free, 1994). These activities are prohibited by the rules which apply to railway shunters, but violation of any one of these rules alone will seldom lead to a fatality, largely because of safeguards built into the system (e.g. adequate space between buffers). Thus 'fatal outcomes are the immediate result of errors committed while in the areas of greater risk ... The shunters between the vehicles slipped ... Those on the line failed to notice the approaching train' and so it is reasonable to conclude (Reason, 1995) that, in some cases at least, 'violations + errors = disaster'.

5.2.5 Unforeseen errors

Unforeseen errors are those in which the occurrence and/or consequences are unfamiliar, unplanned and unexpected. Some knowledge-based mistakes and most of the exceptional violations discussed in the previous sections may be classified as 'unforeseen errors'. Unforeseen errors may also occur as a result of unforeseen events. As was discussed in Chapter 2, unforeseen events in engineering systems may be due to:

1. systems that contain 'complex interactions';
2. new technology and/or information; and
3. unknown physical phenomena.

Even trivial errors may lead to unforeseen consequences; Perrow (1984) cites the following two examples:

1. In 1978 a worker dropped a light bulb while changing the light bulbs in the control room of a nuclear power plant. This created a short circuit in some sensors and controls. The loss of these instruments meant that the operators were not able to monitor the true condition of the plant. Rapid cooling occurred, placing strong internal stresses on the core. No damage was caused, mainly because the reactor core was relatively new.
2. In 1980 a cleaner in a nuclear power plant caught his shirt on a three-inch circuit-breaker, pulled it free, and (unknowingly to him) activated the breaker. This shut off power to the control rod mechanism. Fortunately, the reactor went into automatic shut-down mode.

Finally, it is a natural characteristic of human behaviour to conclude (with the benefit of hindsight) that most 'unforeseen' errors were in fact foreseeable. However, evidence suggests that this is not necessarily the case. For example, in a study of shipping accidents, Wagenaar and Groeneweg (1987) concluded that 'accidents appear to be the result of highly complex coincidences which could rarely be foreseen by the people involved' and 'errors do not look like errors at the time they are perpetrated'. However, in other situations matters may be more complex, with knowledege of past experience being simply forgotten collectively (by a profession, organization etc.). In this case the events would have been unforeseeable to those directly involved but this might not have been the case for those not so involved (Sibly and Walker, 1977).

5.2.6 Error control

Among the possible objectives of a risk analysis may be the study of the effectiveness of various error control programmes. Obviously, the initial stage of error control is the identification of errors and their causes. This is followed by selection of appropriate error control measures. This requires some knowledge of their effectiveness. Error control aims to ameliorate error likely situations and to detect errors. In this context, the influence of Performance Shaping Factors on task reliability is clearly important for error control. Their influence may be assessed by the use of appropriate human reliability databases or through data collected specifically for the tasks of interest (e.g. effect of closer supervision). In this way, the expected benefits (in terms of an increase in task and/or system reliability) of various error control programmes may be compared.

Evidently, system reliability may be increased by the implementation of error-reduction and/or error-tolerant measures (Rouse, 1985). Typically, error-reduction measures are aimed at reducing the overall frequency of errors via external controls (e.g. supervision or checking), training, checklists, alarms, reduction of time pressure, legal sanctions, personnel selection and other measures. Error-tolerance measures on the other hand, are based on the realistic assumption that errors will occur (as an unavoidable consequence of human behaviour) and that it may not be not possible to obtain a completely 'error-free' operating environment. Therefore, error-tolerant measures aim to (1) minimize the opportunity for errors to occur (e.g. utilize aids to reduce task complexity of the physical system), and (2) reduce the consequences if an error does occur (e.g. use of standby and parallel redundant sub-systems).

Industry-specific devices, such as a trip system (which is an automatic protective system) may be able to reduce the possible consequences of errors, since they can shut down the system if a hazardous condition is detected. However, the implementation of these error control measures may themselves lead to situations that may cause (perhaps different) errors to occur. For example, hundreds of alarms went-off within a few seconds at Three Mile Island; thus distracting and probably overloading the control room operators (Senders and Moray, 1991).

An important error control strategy is training. It has been observed by Miller and Swain (1987) that, if training is adequate, internal PSFs have less influence on performance reliability of the individual than do the external PSFs. In terms of error control this is a fortunate observation. External PSFs are generally under the control of the organization; hence human reliability can be improved most effectively by a conscious decision by the organization to implement appropriate error control programmes. These may involve, for example, redesign of control room panels, increased supervision and feedback, and reduction of time pressure.

Internal PSFs (such as stress and inexperience) are also very important as evident from examining the performance of military personnel under combat stress (Miller and Swain, 1987). Reason (1988) cites the following example that took place during the American Civil War: 'during the battle for Atlanta, tree trunks in front of defensive works were found to be bristling with ramrods, fired off prematurely during the loading sequence by troops under attack'. Therefore, it is always useful to remember 'try to change situations, not people' (Kletz, 1991).

Contingency measures also may be implemented to reduce the consequences of human error; for example, a reduction in radiation exposure levels may be achieved if plant personnel and public evacuation procedures can be activated quickly if a severe accident occurs in a nuclear power plant. Clearly, there is a wide range of error control (or risk management) measures that can be applied to a variety of tasks and systems. For further reading, particularly on the general issues, see Reason (1990) and Senders and Moray (1991).

5.3 HUMAN RELIABILITY DATA

An analysis of the probability of failure of a system which attempts to take proper account of human error will require data on the reliability with which humans perform given tasks or actions. Typically, such an analysis requires the following human reliability data for each task or event:

1. error rates: Frequency of error occurrence or Human Error Probability (HEP).

$$HEP = \frac{\text{number of errors}}{\text{total number of opportunities for error}}$$

2. error magnitude: If an error occurs, the error magnitude is the direct consequence of the errors (e.g. is another button activated).
3. error recovery: The rate of detection and correction of errors (e.g. inspection, self-checking).

A human reliability database can only be developed after appropriate human reliability data has been collected and analysed. The sources of human reliability data are now described.

5.3.1 Sources of human reliability data

The event tree logic used in risk analysis can be applied also when human error is incorporated in the analysis. In addition a separate analysis for human error may be performed – the HRA. It has been applied to a range of engineering systems; including nuclear power plants, military and aerospace systems, chemical process plants, structural design and construction etc. The analysis of these systems has led to an extensive amount of human performance data being collected. Despite this, human reliability data in general is still relatively scarce. There is a need to collect more data so that additional errors may be quantified or so that the validity of existing data may be tested. Since human behaviour is a complex phenomenon, the data needs are very large and in consequence the data and techniques used to describe human performance may be subject to considerable uncertainty. Typically, data sources are:

1. monitoring of performance in the workplace,
2. workplace simulators,
3. laboratory studies, and
4. expert opinions.

Human reliability data obtained by direct monitoring of human actions in the workplace (e.g. operators in real plants) is obviously the best and most accurate source of human performance data. However, it may not be possible to directly assess some tasks (e.g. maintenance of equipment, management decision-making). Kirwan (1994) suggests that the lack of availability of workplace human reliability data is due to:

1. difficulties involved in estimating the number of opportunities for error to occur in realistically complex tasks ('the so-called denominator problem');
2. confidentiality;
3. an unwillingness to publish data on poor performance; and
4. a lack of awareness of why it would be useful to collect such data in the first place, and hence a lack of financial resources for data-collection.

Finally, in some industries there may well be employee, trade union, or even employer resistance and/or legislative restrictions on the monitoring of their performance.

The use of workplace simulators and laboratory studies are very useful in providing a controlled environment where the influence of Performance Shaping Factors (PSFs) on human performance may be assessed. However, human performance may be altered if the environment and the tasks do not accurately represent 'real-life'. Using expert opinions to provide quantitative estimates of human performance (e.g. error rates) is an attractive means of data collection, perhaps mainly because this is the least expensive method of obtaining human performance data. Nonetheless, the experts must be carefully selected and their estimates calibrated with known error rate information (see also section 4.3.2). Naturally, the expert assessments are subjective, but for many tasks there is no other feasible method to quantify human performance.

The methodologies for monitoring human performance in the workplace, for using workplace simulators and for conducting laboratory studies, are generally task or industry specific. It is beyond the scope of the present chapter to study these three sources of data in detail; the detailed review of these data sources by Carnino (1986) might be consulted. However, since expert opinions are applied to a variety of industries, it is useful to describe a typical approach, as follows.

For error events in which there is little or no existing human performance data, a technique termed 'Absolute Probability Judgement' (APJ) may be appropriate to provide estimates of error rates. It requires experts to provide a quantitative assessment of error rates (and other human reliability data) with the assessments based on the experience and knowledge of the experts. The APJ technique is appropriate with as few as six expert assessors. Expert opinions may be obtained using one of the following approaches:

1. aggregated individual method,
2. delphi method,
3. nominal group technique or
4. consensus group method (Seaver and Stillwell, 1983).

When non-group consensus methods are used, it is important that the Analysis of Variance (ANOVA) is used to confirm that there is a significant degree of inter-judge consistency between the experts. If the agreement between experts is adequate, then a single point estimate is obtained by statistically aggregating the individual expert assessments (e.g. estimating the geometric mean). There is some empirical support for the APJ method; however, it is unclear how accurate this method is for the estimation of very small error rates. It is sometimes referred to as the 'direct numerical estimation procedure'.

Other techniques also have been developed to use expert opinion to estimate task human error rates; these include:

1. Paired Comparisons (PC),
2. Rankings,
3. Indirect Numerical Estimation, and
4. Multiattribute Utility Measurements (Seaver and Stillwell, 1983).

5.3.2 Human reliability databases

Human reliability databases are appropriately collected, analysed and compiled collections of human reliability data. A large number of databases have been developed (often only applicable to specific industries); some of these will be described in the next section. A human reliability database may be used as a source of reference by risk analysts for use in new (or other) systems. Databases usually contain both generic and specific human performance data. Generic data are applicable to a range of tasks; for example, 'rather simple task performed rapidly or requiring scant attention', and 'complex task requiring high level of compensation and skill' (Williams, 1986). Specific data are only applicable to one particular task (often also restricted to a specific industry); for example, 'making an error of selection in changing or restoring a locally operated valve when the valve to be manipulated is clearly and unambiguously labelled' (Swain and Guttman, 1983).

(a) THERP

Historically, the first reasonably comprehensive, quantifiable human reliability assessment technique was developed by Swain (1963). It was referred to as THERP – Technique for Human Error Rate Prediction. It incorporates both a procedure for formulation and analysis of event tree logic for human error and a database of typical error rates and related PSFs.

Since THERP was developed initially for nuclear power plant applications, the database of task error rates is concerned only with operator procedural errors (Swain, 1963). Later developments led to THERP becoming more comprehensive and the database now provides error rates and recovery factors (i.e. recovery from an error, effect of supervision) for displays, diagnosis of abnormal events, manual controls, locally operated valves, oral instructions and written procedures. The influence of cognitive errors, dependence between events (i.e. error occurrence affected by prior outcome of other task), management and administrative control, stress, staffing, experience levels and other Performance Shaping Factors (PSFs) is also considered as may be seen in the more recent development of the database (Swain and Guttman, 1983) (see also Tables 5.4 and 5.5). The THERP database is comprehensive: it includes 27 tables of human reliability data and data for the following errors, which are particularly relevant for the most common error events encountered in engineering systems:

1. Errors of omission: omits entire task;
                         omits a step in a task; and
2. Errors of commission: incorrect performance of a task;
                         selection error (selects wrong control, issues wrong command);
                         error of sequence;
                         time error (too early, too late);
                         qualitative error (too little, too much).

Error rates and error magnitudes are not necessarily constant, and are likely to vary from individual to individual, with the performance of skilled persons tending to have a lower level of error rates. When considered in terms of a probabilistic distribution of error rates, this suggests that the lognormal distribution is appropriate for modelling human performance data (Swain and Guttman, 1983). Such a distribution implies that it should be possible to estimate the median and variance for the lognormal distribution. Rather than the variance, the meaure of dispersion which has been used is the 'error factor' (EF), which is expressed as

$$\text{EF} = \sqrt{\frac{\Pr(F_{95\text{th}})}{\Pr(F_{5\text{th}})}} \quad (5.1)$$

Table 5.4 Typical operator error rates from the THERP database

Task	BHEP	EF
Errors of omission:		
Omitting a step or important instruction from a formal procedure	0.003	5
Failure to carry out plant policy or scheduled tasks such as periodic checks or maintenance	0.01	5
Failure to use written maintenance procedures	0.3	5
Errors of commission:		
Reading/recording quantitative information from displays		
– Analog meter	0.003	3
– Digital readout (≤4 digits)	0.001	3
– Chart recorder	0.006	3
– Graphs	0.01	3
Recognize that an instrument being read is jammed, if there are no indicators to alert the user	0.1	5
Select wrong control on a panel from an array of similar-appearing controls	0.0005–0.003	3–10
Set a rotary control to an incorrect setting	0.001	10
Operator fails to notice a sticking valve	0.001–0.01	10
Recovery factors:		
Checker will fail to detect errors made by others		
– checking routine tasks, checker using written materials	0.1	5
– checking routine tasks, but without written materials	0.2	5
– checking that involves active participation (e.g. measurements)	0.01	5

Source: adapted from Swain and Guttman (1983).

Table 5.5 Performance shaping factors from the THERP database

	Modifiers for BHEPs	
Stress level	Skilled[a]	Novice[b]
Very low	× 2	× 2
Optimum		
Step-by-step	× 1	× 1
Dynamic (responding to an abnormal event)	× 1	× 2
Moderately high		
Step-by-step	× 2	× 4
Dynamic (responding to an abnormal event)	× 5	× 10
Extremely high (threat stress)		
Step-by-step	× 5	× 10
Dynamic (responding to an abnormal event)	HEP=0.25	HEP=0.5

Notes: [a] at least 6 months' experience in the tasks being assessed.
[b] less than 6 months' experience in the tasks being assessed.
Source: adapted from Swain and Guttman (1983).

Human reliability data

[Figure 5.4 chart: lognormal distribution curve with 5th Percentile and 95th Percentile marked on Error Rate axis]

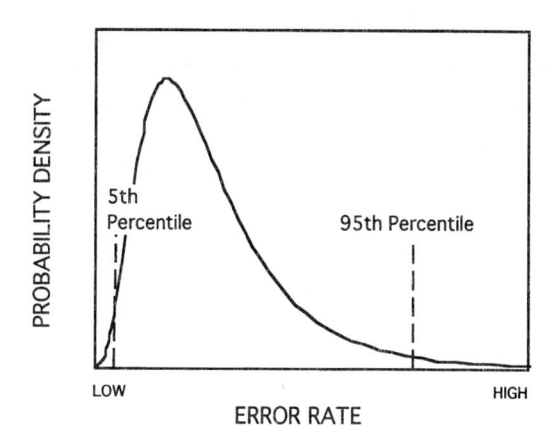

Figure 5.4 Generic lognormal distribution of error rates.

where $Pr(F_{5th})$ and $Pr(F_{95th})$ are the error rates corresponding to the 5th and 95th percentiles respectively of the distribution of error rates, see Figure 5.4. The THERP database lists both the error rate and the error factor for each task. The influence of the error factor on the distribution of error rates is shown in Figure 5.5, for a task with a median error rate of 2×10^{-2}. Other human reliability data such as error magnitudes may be modelled also by the lognormal distribution (e.g. Stewart, 1992).

Despite its comprehensiveness and after nearly three decades of work developing the THERP database, it still does not incorporate all tasks of interest to nuclear power plant risk analysts (Swain and Guttman, 1983). This is not particularly unexpected because of the uniqueness of many

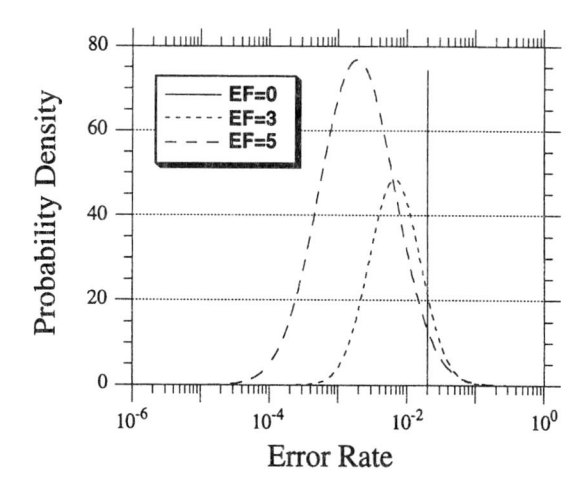

[Figure 5.5 chart: Probability Density vs Error Rate (10^{-6} to 10^0) showing curves for EF=0, EF=3, EF=5]

Figure 5.5 Effect of error factor on distribution of error rates.

tasks and the acknowledgment that the original database and subsequent refinements may not accurately represent all the PSFs affecting human performance. This provides a general indication both of the complexity of nuclear power plant systems and of the difficulty of quantifying relevant human performance characteristics under a range of PSFs.

(b) TESEO

Bello and Colombari (1980) have developed a model to evaluate the error rates of plant control room operators. The model is referred to as TESEO, an empirical technique for estimation of operator errors. It was developed from empirical studies of existing human reliability databases. The operator performance is calculated from the following parameters:

K_1 basic error rate for type of activity to be conducted;
K_2 time available to carry out the activity;
K_3 human operator's characteristics and level of training;
K_4 operator's state of anxiety;
K_5 environmental ergonomics characteristics.

through the following expression:

$$\text{HEP} = K_1\, K_2\, K_3\, K_4\, K_5 \tag{5.2}$$

where HEP is the 'Human Error Probability' or the operator's error rate. Values for each parameter are shown in Table 5.6. Evidently, factors K_2, K_3, K_4 and K_5 are performance shaping factors which modify the generic error rate K_1. For example, if an operator with average experience is to open a remote controlled valve, the control room is noisy and poorly lit and the time to carry out the operation in 20s, then, from Table 5.6, $K_1 = 0.01$, with $K_2 = 0.5$, $K_3 = 1$, $K_4 = 1$ and $K_5 = 10$, resulting in an error rate of 0.05.

(c) HEART

HEART (Human Error Assessment and Reduction Technique) was developed by Williams (1986) as an aid in the estimation of error rates. It uses a list of up to 38 error-producing conditions. There are nine generic basic error rates (BHEP) – defined as the error rate in 'good' conditions – for each error-producing condition i. Each of these has an associated multiplicative factor (K_i) to be applied to the BHEPs – see Table 5.7. The error-producing conditions may be due to, for example, impaired system knowledge, short response time, or poor or ambiguous system feedback. The analyst needs to estimate the proportion of any error-producing condition (p_i) that might exist for a basic task. The overall error rate (HEP) is then calculated, for n error-producing conditions, as

Table 5.6 Performance shaping factors for the TESEO model

ACTIVITY TYPOLOGICAL FACTOR

Type of activity	K_1
Simple routine	0.001
Requiring attention, routine	0.01
Not routine	0.1

TEMPORARY STRESS FACTOR FOR ROUTINE ACTIVITIES

Time available (s)	K_2
2	10
10	1
20	0.5

TEMPORARY STRESS FACTOR FOR NON-ROUTINE ACTIVITIES

Time available (s)	K_2
3	10
30	1
45	0.3
60	0.1

OPERATOR'S TYPOLOGICAL FACTOR

Operator's qualities	K_3
Carefully selected, expert, well trained	0.5
Average knowledge and training	1
Little knowledge, poor training	3

ACTIVITY'S ANXIETY FACTOR

State of anxiety	K_4
Situation of grave emergency	3
Situation of potential emergency	2
Normal situation	1

ACTIVITY'S ERGONOMIC FACTOR

Environmental ergonomic factor	K_5
Excellent microclimate, excellent interface with plant	0.7
Good microclimate, good interface with plant	1
Discrete microclimate, discrete interface with plant	3
Discrete microclimate, poor interface with plant	7
Worst microclimate, poor interface with plant	10

Source: Bello and Colombari, (1980). Reprinted from *Reliability Engineering*, Vol. 1, G.C. Bello and V. Colombari, The human factors in risk analysis of process plants: the control room operator model 'TESEO', 1–14, Copyright 1980, with kind permission from Elsevier Science Ltd, The Boulevard, Langford Lane, Kidlington, OX5 1GB, UK.

Table 5.7 Error-Producing Conditions and their Multiplicative Factor

Error-producing condition	Multiplicative factor (Ki)
A shortage of time available for error detection and correction	× 11
No obvious means of reversing an unintended action	× 8
Poor, ambiguous or ill-matched system feedback	× 4
Operator inexperience (e.g., newly qualified tradesman)	× 3
Little or no independent checking or testing of output	× 3
An incentive to use other more dangerous procedures	× 2
High-level emotional stress	× 1.3
Low workforce morale	× 1.2
Disruption on normal work-sleep cycles	× 1.1

Source: adapted from Williams (1986).

$$\text{HEP} = \text{BHEP} \prod_{i=1}^{n}[1 + p_i(K_i - 1)] \qquad (5.3)$$

A description of remedial measures forms part of HEART. This allows the influence of error control measures to be assessed for each error-producing condition. It should be noted that neither the TESEO nor the HEART techniques have been fully validated, but they do serve as useful guides to quantifying the influence of PSFs. Also, the TESEO and HEART methods do not consider the effect of dependency between tasks.

(d) PROF

PROF (PRediction of Operator Failure rates) is an interactive computer-based database used for the estimation of rates of error in control room tasks, such as operation, calibration, inspection, communication and maintenance tasks (Drager and Soma, 1988). The following PSFs are also included: task complexity, information quality, feedback, breaking of stereotypes, distractions, time pressure, exposure to danger, task relevant experience and personnel qualifications. Human reliability data was obtained from

1. Norsk Hydro and STATOIL control room simulation studies,
2. THERP, and
3. DeSteese et al. (1983).

The operator failure rate (HEP) is obtained from the following regression analysis, based on $k = 1,2,...,m$ PSFs:

$$\log(\text{HEP}) = \sum_{k=1}^{m} \text{PSF}_k\, W_k + C \qquad (5.4)$$

Table 5.8 Typical errors of intention and their HEP's

Errors of intention	HEP UB[a]	HEP LB[b]	EF
Circumvent procedure with potentially catastrophic consequences	7.5E-2	6.0E-5	35
Circumvent procedure with a minor consequence	8.6E-2	3.3E-4	16
Violate procedure and reconfigure equipment	8.3E-2	5.5E-4	12
Violate procedure and devise own formuli	4.7E-2	1.6E-3	5
Symptoms noticed, but incorrect interpretation	1.0E-1	4.2E-3	5
Crews consult inappropriate resources in emergency	1.3E-1	1.9E-3	8
Excessive task duration result in poor judgement	9.0E-2	1.6E-2	2

Note: [a]HEP UB represents a worst case scenario.
[b]HEP LB represents a situation where PSFs have been estimated to be optimal.
Source: adapted from Gertman *et al.* (1992).

where PSF_k is the value of the k-th PSF (on a scale 1 = optimal, 10 = worst), W_k is the weight or importance of the k-th PSFs, and C is a numerical constant. In this technique it is assumed that the PSFs are independent of each other. The user must first select the task being considered, PROF then determines values for W_k for each PSF and for C. Values of PSF_k must be entered by the user. The operator error rate is then calculated from Equation (5.4).

(e) INTENT

The error events considered in the THERP, TESEO and HEART methods are mainly errors of selection and execution (i.e. skill-based slips and rule-based mistakes). However, another important type of error event involves decision-based errors. These are errors of intention, such as where an error occurs as the result of a conscious decision (i.e. knowledge-based mistake or violation) such as where a decision is made to operate outside established procedures. The technique known as INTENT was developed specifically to estimate error rates for errors of intention. Gertman *et al.* (1992) identified 20 potential errors of intention from analyses of nuclear power plant operations. Experts were then used (1) to estimate the 95% upper and lower bounds of a distribution of HEPs (human error probabilities) (see Table 5.8) and (2) to assess the influence of 11 PSFs (performance shaping factors) on the error rate (expressed as PSF 'importance' weights), for each of 20 generic errors of intention. The product of the PSF importance weights and situation specific PSF ratings can then be mapped onto the HEP distribution to produce a site-specific HEP.

(f) TRC

The reliability of plant operators and their responses to accidents is particularly important. In these circumstances time pressure is an important consideration. The problem of interest is also known as time-dependent man-machine interaction. So-called Time Reliability Correlation (TRC) databases have been developed for this situation. They include the influence of diagnostic and decision-making errors (i.e. mistakes). A typical TRC database is the Human Cognitive Reliability (HCR) model (Hannaman and Worledge, 1988), where the non-response probability P(t) is represented by

$$P(t) = \exp - \left\{ \frac{(t/T_{1/2}) - C_{\gamma i}}{C_{\eta i}} \right\}^{\beta i} \tag{5.5}$$

where t is the time available for the crew to complete actions after the occurrence of a stimulus (i.e. onset of emergency). The estimated median time response ($T_{1/2}$) may be obtained from simulator measurements, task analyses or expert opinions. The remaining parameters are influenced by the three types of cognitive processes involved in a response; skill, rule and knowledge-based processes – see also section 5.2.2. The influence of PSFs, such as operator experience, stress level and quality of operator-plant interface, may be included also by modifying the median time parameter. The model has been validated with data from plant simulators. However, the model makes no attempt to quantify the dependency between tasks and is not designed for procedural events in which no time pressure is imposed. Other TRC databases have also been developed (see Dougherty and Fragola, 1988).

(g) SLIM

The Success Likelihood Index Methodology (SLIM) is an interactive computer-based technique based on expert opinion (Embry et al., 1984). For each task, an expert identifies the relevant PSFs and estimates the effect of each PSF on the likelihood of task success (i.e. on task performance). The relative likelihood of successful task completion (relative HEP) can then be derived. In order to establish an estimate of (absolute) HEP, a calibration needs to be performed. For this it is necessary that the HEPs for (at least) two tasks be known; generally such data is taken from the THERP database. The relative likelihoods can be transformed in this way into HEPs; usually a logarithmic calibration relationship is employed. Note that the derived HEPs are sensitive to the choice of calibration tasks, and that it is not generally accepted that the use of the logarithmic calibration relationship is valid.

(h) MAPPS

MAPPS (Maintenance and Personal Performance Simulation) is a computer-based simulation model developed by Siegel et al. (1984). It provides task reliability estimates for maintenance performance, including corrective, preventative and diagnostic maintenance tasks. The user may input information on task variables, PSFs (in this case factors such as heat, stress and fatigue), error recovery factors and other parameters which may affect task performance. These parameters are then quantified in terms of their influence on task reliability using an algorithmic model. The average probability of success and the estimated task completion time are obtained through Monte Carlo computer simulation.

(i) Other human reliability databases

Other human reliability databases have also been developed; these include:

1. AIR:	American Institute for Research (AIR) Data Store was developed to assess specific design features for electronic equipment operability (Munger et al., 1962).
2. Aerojet General Method:	Irwin e al. (1964) expanded the AIR Data Store to include maintenance and inspection tasks associated with the Titan II propulsion system.
3. Bunker-Ramo Tables:	The AIR Data Store was further expanded to include data from 37 different experimental studies of operator performance (Hornyak, 1967).
4. OPREDS:	The Operational Recording and Data System (OPREDS) contains data on switch-turning and button-pressing obtained from automatic monitoring of human operators in US Navy ships (Urmston, 1976).
5. ASRS:	The Aviation Safety Reporting System (ASRS) contains civil aircraft incident/accident data obtained from voluntary reporting (Federal Aviation Administration, 1979).
6. SRL:	The Savannah River Laboratory have developed a database of operator error rates and equipment failure rates for the

| | operation of nuclear-fuel reprocessing plants (Durant et al., 1988).
7. NUCLARR: | The Nuclear Computerised Library for Assessing Reactor Reliability (NUCLARR) contains error rates and equipment failure rate data for Nuclear Power Plant operator and maintenance tasks (Gertman et al., 1990).

Extensive reviews of existing human reliability databases have been given by Topmiller *et al.* (1982); Dhillon (1986, 1990); Miller and Swain (1987); Dougherty and Fragola (1988); Embry and Lucas (1989); Kirwan (1994) and others. Other databases have been proposed (e.g. Sandia Human Error Rate Bank – Rigby, 1967) but have never been established; this is partly due to the reluctance of organizations to provide data on their own errors (Meister, 1978).

(j) Existing literature

It appears that most of the existing databases are concerned mainly with human actions associated with the operation and maintenance of nuclear, chemical or process plants. This is not surprising since it is often a regulatory requirement that these industries conduct comprehensive quantitative/probabilistic risk analyses (including the effect of human error). However, the relative lack of databases for other systems does not necessarily imply that human reliability data do not exist for these systems. For example, human reliability data often are obtained from experimental studies described in the literature, including journals, conference proceedings, and research reports. The following list contains some tasks for which human reliability data are available:

1. computations – calculations, table look-up, numerical ranking (Melchers, 1984);
2. checking of design computations (Stewart and Melchers, 1989);
3. chart look-up (Beeby and Taylor, 1973);
4. assembly of electronic equipment (Rigby and Swain, 1968);
5. computer software (Lipow, 1982; Gondran, 1986);
6. operator actions in chemical process plants (Kletz, 1991);
7. reinforced concrete construction (Stewart, 1993); and
8. pilot and aircrew performance during emergency situations (Rigby and Edelman, 1968).

Some typical error rates for these tasks are presented in Table 5.9. A useful reference for assessing human reliabilities for a wide range of psychomotor and cognitive tasks is the Engineering Data Compendium (Boff and Lincoln, 1988).

Human reliability data

Table 5.9 Human error rates for various tasks

	Error rate	Source
Design computations		
Chart look-up	0.020	Beeby and Taylor (1973)
Table look-up	0.013	Melchers (1984)
Onestep calculation (e.g., a×b)	0.013	Melchers (1984)
Twostep calculation (e.g., a×b−c)	0.026	Melchers (1984)
Self-checking of calculations	0.90[a]	Stewart and Melchers (1989)
Independent checking of calculations	0.65[a]	Stewart and Melchers (1989)
Reinforced concrete construction		
Reduced number of reinforcing bars	0.022	Stewart (1993)
Increased number of reinforcing bars	0.011	Stewart (1993)
Pilot and aircrew performance[b]		
Fire in cabin during takeoff	0.23	Rigby and Edelman (1968)
Landing gear fails to extend	0.14	Rigby and Edelman (1968)
Loss of visual contact with ground	0.02	Rigby and Edelman (1968)
Electronic component assembly		
Wires reversed	0.001–0.005	Rigby and Swain (1968)
Component reversed	0.002	Rigby and Swain (1968)
Component omitted	0.00003	Rigby and Swain (1968)
Insufficient solder	0.0009	Rigby and Swain (1968)
Computer software		
Number of "bugs" per line of code	0.002	Lipow (1982)
Significant error in applications software	0.01–0.001	Gondran (1986)

[a] error rate refers to the probability of not correcting an incorrect result.
[b] error rate refers to error rate per individual task.

5.3.3 Validation of human reliability databases

In order to estimate the validity of human reliability databases, Poucet (1988) conducted a benchmark exercise to assess analyst error rate predictions for two 'well defined' nuclear power plant operator tasks; namely (1) non-detection of a failure in a three-part free-flow check valve and (2) manual valves in a venting line being left open after a test. Fifteen teams of analysts from 11 countries participated in the exercise. The analysts' assessments were obtained using to one or more of the following human reliability databases: THERP, SLIM, HEART and MAPPS. Poucet (1988) showed that the quantitative results (i.e. error rates) obtained from THERP were subject to the largest analyst-to-analyst variability. However, there was shown to be general agreement (to within an order of magnitude) in the error rates obtained from the four databases. This variation is not unexpected since the quantification of error rates requires subjective judgements about PSF ratings, dependency etc. The study concluded that the analyst-to-analyst variability was due mainly to the type of database used. However analyst-to-analyst variability was not large when all analysts use the same database.

Few other human reliability validation studies have been conducted; two of these are described by Williams (1985) and Kirwan (1994). One study considered the AIR data store, SLIM, APJ, PC, Technique for Establishment of Personnel Performance Standards (Smith et al., 1969), and digital simulation (Siegel and Wolf, 1969). The validations consisted of comparing human reliability assessments with the known probability for an event. Assessments obtained from the APJ method were found to be accurate to within a factor of ten; this is in agreement with the suggestion by Williams (1985) that the predictive precision of human reliability estimates should be within an order of magnitude. The other sources of human reliability data showed larger variations from the known probabilities. This led to the suggestion that the APJ method is the 'least worst' of all the methods, a conclusion in broad agreement with a more recent study conducted by Kirwan (1988). It was also observed that human reliabilities in excess of 10^{-1} and below 10^{-4} are the most difficult to predict with any accuracy and that there can be no substitute for the collection and analysis of 'hard data' (i.e. data obtained from direct observation) (Williams, 1985).

5.4 SUMMARY

It has been observed already that reliability analyses involving human error and based on event tree logic and human reliability databases are used by nearly all human reliability analysts. However, it should be recognized that event tree logic is inherently restricted to task events for which errors can be quantified. This requirement for quantification has meant that most sources of human reliability data only consider the incidence and control of active errors (generally skill-based slips and rule-based mistakes) and that therefore the human reliability databases described in this chapter are mainly concerned with operator (e.g. control room) errors. However, even these databases do not contain error data for every possible human activity in a control room. It follows that the choice of human reliability data from human reliability databases requires a subjective judgement by the analyst about the task itself, its PSF ratings, the degree of dependency which may be involved, and other factors (Poucet, 1988).

An important issue is that the above approach generally does not include errors involving diagnosis or high-level decision-making (i.e. knowledge-based mistakes). There is also a continuing need to obtain human reliability information on the influence of

1. latent errors (e.g. influence of managerial decisions such as reduction in maintenance staff),

2. violations, and
3. some error control measures (e.g. legal sanctions, personnel selection).

The source of such quantitative information is as yet unclear and the omission of errors of this type from most risk analyses may be a matter of some significance.

The fact that, by definition, HRAs can not include the effects of 'unforeseen' events should also be considered in an analysis. However, it would be expected that the incidence of 'unforeseen' events will decrease as the operational experience of the system increases with time. This means also that HRAs can be revised as new initiating events or consequences are identified. Despite these deficiencies, HRAs are particularly useful for assessing the comparative effectiveness of various error control programmes. The event tree logic used in HRA also produces a structured, logical record of possible error events. This process may itself aid the identification of new or unexpected initiating events (sequences or combinations) or event consequences that otherwise might have been overlooked in a risk analysis.

Finally, the uncertainty associated with modelling human actions has led Williams (1985) to conclude that 'The developers of human reliability assessment techniques have yet to demonstrate, in any comprehensive fashion, that their methods possess much conceptual, let alone empirical, validity.' This is perhaps a not unreasonable summation of the current state of affairs in human reliability assessments.

REFERENCES

Bea, R.G. (1989) *Human and Organizational Error in Reliability of Coastal and Offshore Platforms*, Civil College Eminent Overseas Speaker Programme, Institution of Engineers, Australia.

Beeby, A.W. and Taylor, H.P.J. (1973) How well can we use graphs, *The Communicator of Scientific and Technical Information*, **17**, October, 7–11.

Bello, G.C. and Colombari, V. (1980) The human factors in risk analysis of process plants: the control room operator model 'TESEO', *Reliability Engineering*, **1**, 1–14.

Boff, K.R. and Lincoln, J.E. (1988) *Engineering Data Compendium: Human Perception and Performance*, Harry G. Armstrong Aerospace Medical Research Laboratory, Wright-Patterson Air Force Base, Ohio.

Carnino, A. (1986) Role of data and judgement in modelling human errors, *Nuclear Engineering and Design*, **93**, 303–9.

Christensen, J.M. and Howard, J.M. (1981) Field experience in maintenance. In J. Rasmussen and W.B. Rouse (eds), *Human Detection and Diagnosis of System Failures*, Plenum Press, New York, 111–33.

DeSteese, J.G. et al. (1983) *Human Factors Affecting the Reliability and Safety of LNG Facilities: Control Panel Design Enhancement*, GRI-81/0106.

Dhillon, B.S. (1986) *Human Reliability with Human Factors*, Pergamon Press, New York.

Dhillon, B.S. (1990) Human error data banks, *Microelectronics and Reliability*, **30**(5), 963–71.

Dougherty, E.M. and Fragola, J.R. (1988) *Human Reliability Analysis*, Wiley, New York.

Drager, K.H. and Soma, H.S. (1988) PROF: a computer code for prediction of operator failure rate. In B.A. Sayers (ed.), *Human Factors and Decision Making*, Elsevier Applied Science, 158–69.

Durant, W.S., Lux, C.R. and Galloway, W.D. (1988) Data bank for probabilistic risk-assessment of nuclear-fuel reprocessing plants, *IEEE Transactions on Reliability*, **37**(2), 138–42.

Embry, D.E., Humphreys, P., Rosa, E.A. *et al.* (1984) *SLIM-MAUD An Approach to Assessing Human Error Probabilities Using Structured Expert Judgement*, NUREG/CR-3518, US Nuclear Regulatory Commission, Washington, DC.

Embry, D.E. and Lucas, D.A. (1989) Human reliability assessment and probabilistic risk assessment. In V. Colombari (ed.), *Reliability Data Collection and Use in Risk and Availability Assessment*, Springer-Verlag, Berlin, 343–57.

Federal Aviation Administration (1979) *Aviation Safety Reporting Programs*, FAA Advisory Circular 00-46B, Washington, DC.

Finnegan, J., Rau, C.A., Rettig, T. and Weiss, J. (1980) Personnel errors and power plant reliability, *Proc. Ann. Reliability and Maintainability Symposium*, 290–7.

Free, R. (1994) *The Role of Procedural Violations in Railway Accidents*, PhD thesis, University of Manchester.

Gardenier, J.S. (1981) Ship navigational failure detection and diagnosis. In J. Rasmussen and W.B. Rouse (eds), *Human Detection and Diagnosis of System Failures*, Plenum Press, New York, 49–74.

Gertman, D.I., Gilmore, W.E., Galyean, W.J. *et al.* (1990) *Nuclear Computerised Library for Assessing Reactor Reliability (NUCLARR), Vol. 1: Summary Description and Vol. 5: Data Manual*, NUREG/CR 4639, US Nuclear Regulatory Commission, Washington, DC.

Gertman, D.I., Blackman, H.S., Haney, L.N. *et al.* (1992) INTENT: a method for estimating human error probabilities for decision-based errors, *Reliability Engineering and System Safety*, **35**, 127–36.

Gondran, M. (1986) Launch meeting of the European Safety and Reliability Association, Brussels (see Kletz, 1991).

Hannaman, G.W. and Worledge, D.H. (1988) Some developments in human reliability analysis approaches and tools, *Reliability Engineering and System Safety*, **22**, 235–56.

Hayashi, Y. (1985) Hazard analysis in chemical complexes in Japan – especially those caused by human error, *Ergonomics*, **28**(6), 835–41.

Hollnagel, E. (1993) *Human Reliability Analysis: Context and Control*, Academic, London.

Hornyak, S.J. (1967) *Effectiveness of Display Subsystems Measurement and Prediction Techniques*, RADC Report TR-67-292, Rome Air Development Centre, Griffiss Air Force Base, New York.

Ingles, O.G. (1979) Human factors and error in civil engineering. In *Third International Conference on Statistics and Probability in Soil and Structural Engineering*, Sydney, Australia, 402–17.

Irwin, I.A., Levitz, J.J. and Freed, A.M. (1964) *Human Reliability in the Performance of Maintenance*, Report LRP 317/TDR-63-218, Aerojet General Corporation, Sacramento, CA.

Kinney, G.C., Spahn, M.J. and Amato, R.A. (1977) *The Human Element in Air Traffic Control: Observation and Analysis of Performance of Controllers and Supervisors in Providing Air Traffic Control Separation Services*, Report No. MTR-7655, METREK Division, MITRE Corporation.

Kirwan, B. (1988) A comparative evaluation of five human reliability assessment techniques. In B.A. Sayers (ed.), *Human Factors and Decision Making*, Elsevier Applied Science, 87–109.
Kirwan, B. (1994) *A Guide to Practical Human Reliability Assessment*, Taylor & Francis, London.
Kletz, T. (1991) *An Engineer's View of Human Error*, Institution of Chemical Engineers, Rugby, Warwickshire.
Lipow, M. (1982) Number of faults per line of code, *IEEE Transactions on Software Engineering*, **SE-8**(4), 437–9.
Loss, J. and Kennett, E. (1987) *Identification of Performance Failures in Large Structures and Buildings*, School of Architecture and Architecture and Engineering Performance Information Center, University of Maryland.
Matousek, M. and Schneider, J. (1977) *Untersuchungen zur Struktur des Sicherheitsproblems bei Bauwerken*, Report No. 59, Institute of Structural Engineering, Swiss Federal Institute of Technology, Zurich, 1977. (See also Hauser, R. (1979) Lessons from European failures, *Concrete International*, 21–5.)
Meister, D. (1978) Subjective data in human reliability estimates. In *Proceedings of the 1978 Annual Reliability and Maintainability Symposium*, IEEE, 380–4.
Melchers, R.E. (1984) Human error in structural reliability assessments, *Reliability Engineering*, **7**, 61–75.
Miller, D.P. and Swain, A.D. (1987) Human error and human reliability. In G. Salvendy (ed.), *Handbook of Human Factors*, Wiley, New York, 219–50.
Munger, S.J., Smith, R.W. and Payne, D. (1962) *An Index of Electronic Equipment Operability: Data Store*, Report AIR-C43-1/62 RP(1), American Institute for Research, Pittsburgh.
Norman, D.A. (1981) Categorisation of action slips, *Psychological Review*, **88**, 1–15.
Paté-Cornell, M.E (1989) Organizational control of system reliability – a probabilistic approach with application to the design offshore platforms, *Control-Theory and Advanced Technology*, **5**(4), 549–68.
Perrow, C. (1982) The President's Commission and the normal accident. In C.P. Wolf and V.B. Shelanski (eds), *Accident at Three Mile Island: The Human Dimensions*, D.L. Sills, Westview Press, Colorado, 173–84.
Perrow, C. (1984) *Normal Accidents: Living with High-Risk Technologies*, Basic Book Inc., New York.
Poucet, A. (1988) Survey of methods used to assess human reliability in the human factors reliability benchmark exercise, *Reliability Engineering and System Safety*, **22**, 257–68.
Rasmussen, J. (1979) *What Can Be Learned From Human Error Reports*, Riso National Laboratory, Denmark, Riso Report n-17-79.
Rasmussen, J., Duncan, K. and Leplat, J. (1987) *New Technology and Human Error*, Wiley, Chicester.
Reason, J. (1988), Stress and cognitive failure. In S. Fisher and J. Reason (eds), *Handbook of Life Stress, Cognition and Health*, Wiley, New York, 405–21.
Reason, J. (1990) *Human Error*, Cambridge University Press, Cambridge.
Reason, J. (1995) A systems approach to organizational error, *Ergonomics*, **38**(8), 1708–21.
Rigby, L.V. (1967) *The Sandia Human Error Rate Bank*, Report No. SC-R-67-1150, Sandia Laboratories, Albuquerque, New Mexico.
Rigby, L.V. (1970) The nature of human error, *Annual Technical Conference of the ASQC*, American Society for Quality Control, Milwaukee, Wisconsin, 457–66.
Rigby, L.V. and Edelman, D.A. (1968) A predictive scale of aircraft emergencies, *Human Factors*, **10**(5), 475–82.

Rigby, L.V. and Swain, A.D. (1968) Effects of assembly error on product acceptability and reliability, *Proceedings of the Seventh Annual Reliability and Maintainability Conference*, ASME, New York, 312–19.

Rouse, W.B. (1985) Optimal allocation of system development resources to reduce and/or tolerate human error, *IEEE Transactions on Systems, Man, and Cybernetics*, **SMC-15**(5), 620–30.

Scott, R.L. and Gallaher, R.B. (1979) *Operating Experience with Valves in Light-Water-Reactor Nuclear Power Plants for the Period 1965–1978*, Report No. NUREG/CR-0848, US Nuclear Regulatory Commission, Washington, DC.

Seaver, D.A. and Stillwell, W.G. (1983) *Procedure for Using Expert Judgement to Estimate Human Error Probabilities in Nuclear Power Plant Operations*, NUREG/CR-2743, US Nuclear Regulatory Commission, Washington, DC.

Senders, J.W. and Moray, N.P. (1991) *Human Error: Cause, Prediction and Reduction*, Lawrence Erlbaum Associates, New Jersey.

Sheppard, B. (1992) Hazardous waste: management and disposal. In R.F. Cox (ed.), *Assessment and Control of Risks to the Environment, and to People*, The Safety and Reliability Society, Manchester, 2/1–2/12.

Sibly, P.G. and Walker, A.C. (1977) Structural accidents and their causes, *Proc. Inst. Civil Engrs*, **62**(1), 191–208.

Siegel, A.I. and Wolf, J.J. (1969) *Man-Machine Simulation Models: Performance and Psychological Interactions*, Wiley, New York.

Siegel, A., Bartter, W.D.., Wolf, J.J. et al. (1984) *Maintenance Personnel Performance Simulation (MAPPS) Model: Description of Model Content, Structure and Sensitivity Testing*, NUREG/CR-3626, US Nuclear Regulatory Commission, Washington, DC.

Smith, R.L., Westland, R.A. and Blanchard, R.E. (1969) *Techniques for Establishing Personnel Performance Standards (TEPPS): Results of Navy User Test*, Report PTB-70-5, Vol. III, Personnel Research Division, Bureau of Navy Personnel, Washington, DC.

Stewart, M.G. (1992) Simulation of human error in reinforced concrete design, *Research in Engineering Design*, **4**(1), 51–60.

Stewart, M.G. (1993) Structural reliability and error control in reinforced concrete design and construction, *Structural Safety*, **12**, 277–92.

Stewart, M.G. and Melchers, R.E. (1989) Checking models in structural design, *Journal of Structural Engineering*, ASCE, **115**(6), 1309–24.

Swain, A.D. (1963) *A Method for Performing a Human Factors Reliability Analysis*, Monograph SCR-685, Sandia National Laboratories, Albuquerque, New Mexico

Swain, A.D. and Guttman, H.E. (1983) *Handbook of Human Reliability Analysis with Emphasis on Nuclear Power Plant Applications*, NUREG/CR-1278, US Nuclear Regulatory Commission, Washington, DC.

Topmiller, D.A., Eckel, J.S. and Kozinsky, E.J. (1982) *Human Reliability Data Bank for Nuclear Power Plant Operations: A Review of Existing Human Reliability Data Banks*, NUREG/CR 2744, US Nuclear Regulatory Commission, Washington, DC.

Turner, B.A. (1978) *Man-made Disasters*, Wykeham Publications, London.

Ujita, H. (1985) Human error classification and analysis in nuclear power plants, *Journal of Nuclear Science and Technology*, **22**(6), 496–598.

Urmston, R. (1976) *Operational Performance Recording and Evaluation Data System (OPREDS)*, Descriptive Brochures, Code 34400, Navy Electronics Laboratory Center, San Diego.

Wagenaar, W.A. and Groeneweg, J. (1987) Accidents at sea: multiple causes and impossible consequences, *International Journal of Man-Machine Studies*, **27**, 587–98.

Williams, J.C. (1985) Validation of human reliability techniques, *Reliability Engineering*, **11**, 149–62.

Williams, J.C. (1986) HEART – a proposed method for assessing and reducing human error, *Ninth Advances in Reliability Technology Symposium*, Bradford, England, pp. B3/R/1 – B3/R/13.

Wood, D.D. (1984) *Some Results on Operator Performance in Emergency Events*, Institute of Chemical Engineers Symposium Series, 21–31.

CHAPTER 6

System evaluation

6.1 INTRODUCTION

This chapter will focus on the evaluation of the probability of system failure. The first part of the chapter will concentrate on the system evaluation of event and fault trees, as was introduced in Chapter 3. It will be assumed that there is sufficient information available about probabilities of failure for the individual components and other system elements making up event or fault trees. These data input matters were discussed in Chapters 4 and 5.

As already noted in section 4.4, for some sub-systems it is possible, or necessary, to estimate the probability of failure from information about loads and resistances (or demands and capacities), with these quantities described in a probabilistic manner. As noted there, it could be considered that such an approach is more fundamental than one relying only on observed failure rate data. This would allow predictive analysis of the physics of the system elements or components, particularly if the physics is properly understood; since then the parameters of interest will be known as functions of particular environmental conditions, such as temperature and pressure. The analyst could then predict accurately how the environment might affect the failure modes and the failure rate of the sub-system (and hence the system itself) (Shooman, 1968). Unfortunately, the lack of such understanding and of models usually means that analysts must use observational data, relying on experience and intuition in justifying its application. As noted in Chapters 4 and 5, in some cases the estimated probabilities of occurrence will be very uncertain due to the lack of appropriately sharp data.

When the loadings or demands and the resistance or capacity are best described by random variables or by continuous stochastic processes, the analysis for the probability of failure introduced in section 4.4 applies. However, in practical situations it rapidly becomes more complex than discussed there. The integration of the physical modelling of components, sub-systems and other system elements into more conventional system risk analysis was outlined by Shooman (1968) in an electrical engineering oriented reliability text. More recent discussions, in more wide-ranging frameworks, with a somewhat more

Frequentist data, probabilities and uncertainties   155

practical emphasis, are available (e.g. Cornell, 1982; Der Kiureghian and Moghtaderi-Zadeh, 1982) but appear to have had little practical application (with the possible exception within the nuclear industry). Generally similar comments apply about the effect of time, such as when the system wears out, or when the system is subject to stochastic processes such as arise from wind, wave or earthquake loads. How to deal with these matters is the aim of the second part of this chapter (section 6.8) where more detailed reliability theory is given than is usually discussed in texts dealing with traditional risk analysis.

System evaluation methods fall into three broad categories:

1. qualitative analyses in which numbers and probabilities are not used extensively or at all;
2. Quantified Risk Analysis (QRA) in which system element performance (or event outcomes) and system risk are given as numerical point estimates; and
3. Probabilistic Risk Analysis (PRA) in which system element performance is given as a random variable, so that the variability and the uncertainty of the variables is propagated through the analysis, leading to the system risk being represented as a probability distribution (i.e. it is a dependent random variable).

Each of these methods are outlined in the present chapter.

Human Reliability Analysis (HRA) is essentially a QRA or PRA with human reliability data used in a manner similar to that of system element performance data (see also Chapter 5). It follows that system evaluation techniques using HRA models are essentially identical to those for QRA and PRA methods. (Of course, qualitative methods can also be used.) As already noted repeatedly, human error usually will add significantly to the overall system risk. For this reason it is becoming more usual to consider the effect of human error directly in a risk analysis, or, as is often the case in the nuclear industry, to perform an entirely separate HRA. Useful overviews have been given for nuclear, chemical process and offshore facilities (Kirwin, 1994), for structural design and construction tasks (Stewart, 1993), for component assembly (Rigby and Swain, 1968) and for software development (Lipow, 1982). Further details of HRA methods are available from Dougherty and Fragola (1988), Hollnagel (1993) and Kirwin (1994).

6.2 FREQUENTIST DATA, PROBABILITIES AND UNCERTAINTIES

In Chapter 3 it was observed that components, items of equipment and some other system elements are discrete events which have outcomes which either occur or do not (see Figure 3.6). In some cases there may be more than two possible outcomes – the event occurs, it does not or there is some undecided situation (for example, see Figure 3.5). Probability estimates for the outcomes associated with an event need to be available for a risk analysis to be conducted. Chapters 3 and 4 dealt with the historical bases for data. Such data is of a frequentist nature and reference is made to 'frequency of occurrence' as a description based on previous relevant experience. It is often assumed that such data may be used for the event outcome 'probabilities' for the event(s) in the system being studied. This means that the past data from other systems, usually, is assumed to be translatable into an expectation of what might be the case for the system under study. To do so makes the assumption that the event being considered is of a closely similar class to the events making up the past experience. It also assumes that extrapolation from past experience to the present situation is appropriate and valid. In many cases, of course, it will be obvious when these requirements are fulfilled, but care is required nevertheless. As also noted in Chapter 4, the distinction between frequentist data and probability inference based on that data is an important one (e.g. Kaplan and Garrick, 1981). Moreover, the probability estimate, or more generally the estimate of the probability density function, is often a subjective probability assessment (e.g. Apostolakis, 1987).

As will have become evident from the discussion in the previous chapters, the probability associated with the outcome of an event is always a conditional probability. This means that the probability of a particular outcome for an event is dependent on the conditions at the time the event is considered. Sometimes in risk analysis work this aspect is forgotten or ignored, with the probability estimate for the event outcome being obtained, by inference, from the frequency of similar components or events, without too much notice being taken of the precise ambient or operating or other conditions. When such information is known, however, it should be considered, and this can be done in a simple manner.

For example, the success rate of a valve operating correctly in a particular application may be temperature dependent. To obtain an estimate from existing information about the likely success rate for the operation of the valve, one approach would be to select the success rate corresponding to the mean temperature over the service life of the valve. However, this approach could be considerably in error if the success rate is not a linear function of temperature. In general it is much better to obtain a weighted average, obtained by integrating the temperature specific valve characteristic $C(\theta)$ over the probability distribution $f_\theta(\theta)$ of the temperature distribution, so as to obtain the mean valve characteristic C, thus:

$$C = \int_\theta C(\theta).f_\theta(\theta)d\theta \tag{6.1}$$

where $f_\theta(\theta)$ is the frequency distribution for the temperature. Evidently, (6.1) will produce a better estimate of the valve mean operating success rate than some value based, perhaps, on the mean temperature.

However, this approach does require the temperature dependent operating characteristics of the valve to be known.

A generally similar approach should be used whenever the event outcome is subject to a matter about which there is uncertainty and which could be taken out of consideration in an analogous way. For example, as already mentioned in section 4.3.3, observational, historical data can be supplemented by subjective information, such as expert opinions, using Bayes Theorem, with (Bayesian) estimates for $f_\theta(\theta)$. The approach is then the same as given in Equation (6.1) above but now with Bayesian frequency estimates for $f_\theta(\theta)$.

Irrespective of the source of the estimates, it might be expected that, in general, there will be some degree of uncertainty in the estimated probabilities (expected rate of occurrence) for describing the performance of system elements (i.e. outcomes of events), as discussed in section 3.5.2. In more conventional risk analysis, information about event outcome uncertainty (1) is not readily available and (2) is often ignored. If it is available, then it is commonly employed as a guide to making a conservative estimate for the rate of occurrence. This might be a percentile value, if the probability information is good, or simply a conservative (and deterministic) 'point estimate value'. The latter approach is typical of QRA procedures (section 6.4).

A better treatment to event outcome uncertainty in event and fault tree analysis is to use uncertainty information (such as the variance or the standard deviation) directly for each of the events. This can be done by using probabilistic methods for the analysis. This is the topic of section 6.5, where PRA procedures are discussed.

6.3 QUALITATIVE RISK ANALYSIS

The simplest approach to risk analysis is to use subjective assessments of risks and to rank these in a subjective manner. The top ranking risk items are then those considered to require attention, such as through amelioration techniques, risk management approaches or through a more thorough risk analysis to determine whether risk acceptance criteria are likely to be exceeded.

Expert opinions are critical in analyses of this type. The experts must understand both the consequences and the likelihood of occurrence of the consequences and then rank each according to the estimated effect and the estimated magnitude. These estimates must then be combined to obtain an estimate for the risk associated with each identified consequence. Where a number of experts are consulted, consensus may be difficult to obtain about the effects of a (proposed) facility, the hazards likely to be associated with it and the possible consequences of failure and the associated probabilities of occurrence of the consequences. In this case (which is probably the usual situation), meetings at which the experts can discuss the issues may be appropriate in an attempt to develop a consensus opinion, or implementation of the Delphi (or similar consensus-seeking) technique may be necessary to obtain a degree of consistency among the expert opinions.

In some cases a magnitude rating for risk may not be possible and resort may have to be had to a simple ranking, in which it is agreed only that the risk (A) > risk(B) > risk(C), etc. without judgement about their relative magnitudes.

A refinement of this process is to keep consequences and probability of occurrence of the consequences separate. Each is assessed using a qualitative descriptor, such as 'high', 'medium', 'low' etc. and the results entered into a 'likelihood-consequence' table, see Table 6.1. A further subjective judgement is then desirable to assess each of the matrix entries (see shaded zones in Table 6.1).

The main problem with qualitative assessments is that there is little indication on an absolute scale how serious the risk might be, particularly for comparison to other risk sources. It is not, therefore, very useful for making rational decisions, although it might be of use in preliminary analyses. It is sometimes attempted to put numbers to the subjective assessments, such as shown at the left and top of Table 6.1 and to use these to obtain risk estimates for each cell in the table (by multiplying likelihood by consequence). It should be clear that use of the numbers so obtained for comparative purposes implies a value-system of undetermined characteristics. For this reason, the relative ranking of risks could be highly misleading. A better way of analysing risks is to turn to a quantitative method.

6.4 QUANTIFIED RISK ANALYSIS (QRA)

A risk analysis in which the probability of occurrence associated with each event is simply a number, will be termed herein a 'point estimate' (or deterministic) analysis. This is the approach adopted for Quantified Risk Analysis (QRA) such as is common in the chemical engineering industry (ISG, 1985). Typically the numbers describing the event outcome probabilities will be values considered to be suitably conservative in terms of evaluating system risk. Such numbers have been termed, rather accurately, 'Conservative Best Estimates', reflecting the fact that

1. they are the best that can be obtained given the available information,
2. they are estimates rather than precisely determined quantities, and
3. they are considered to be 'conservative' in terms of the overall system analysis outcomes.

Table 6.1 Likelihood–consequence matrix

Likelihood	Consequences				
	Insignificant [1]	Minor [2]	Moderate [3]	Major [4]	Catastrophic [5]
Almost certain [5]	High	High	High	High	High
Likely [4]	Significant	High	High	High	High
Moderate [3]	Low	Moderate	High	High	High
Unlikely [2]	Low	Low	Moderate	High	High
Rare [1]	Low	Low	Moderate	High	High

Legend: High risk, Moderate risk, Significant risk, Low risk

The consequences can be classified as follows (see also AS/NZS 4360:1995):

Catastrophic   Widespread deaths, long-term health effects, off-site toxic release, extreme financial loss.
Major   Some deaths, extensive injuries, short-term health effects, major financial loss.
Moderate   Injuries with medical intervention required, on-site release, high financial loss.
Minor   Minor injuries with first aid treatment required, no significant release of chemicals, some financial loss.
Insignificant   No injuries, low financial loss.

It should be clear that while using conservative best estimates for individual event outcomes may be quite justifiable, they impart no information to the analyst about the uncertainty inherent in the system outcome. The reason for this is that it is not possible to estimate the combined effect of all the various conservative assumptions embedded in the analysis. This aspect is handled better by system error propagation analysis (PRA).

6.4.1 Event trees

A discussion of event trees and their function in system analysis was given in section 3.4.3. Recall that they describe the sequence of events

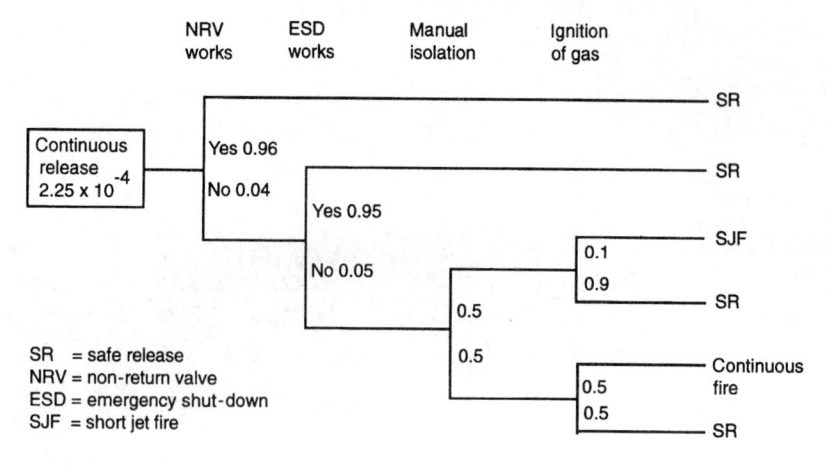

Figure 6.1 Event tree.

from some starting (or initial) event to outcomes. Usually these are events for which there are associated consequences.

Consider the example event tree shown in Figure 6.1. For each branch of the event tree, that is, for each of the possible outcomes which can be associated with the event, the probabilities associated with each outcome must sum to unity (assuming, as always, that one outcome will be one in which nothing happens): $\sum_i p_i = 1$.

Thus for event E_a in Figure 6.1:

$$P(\text{event occurs}) + P(\text{event does not occur}) = 1 \quad (6.2)$$

The outcome probability $P(output)$ for a branch such as branch 'i' will be the combination of the probability of the event occurring $P(y)$ and the probability $P(input)$ that the sequence of previous events actually led up to this particular event, hence

$$P(output) = P(input) \cap P(y) \quad (6.3)$$

where $A \cap B$ represent the intersection of the two probabilities. If the events (input) and (y) are independent events, there results:

$$P(output) = P(input).P(y) \quad (6.4)$$

By extending the argument from the initial event E_i to the outcome event E_o, then

$$p_{out} = P(E_i \cap E_a \cap E_b) \quad (6.5)$$

and if the events in the sequence are all independent, this becomes

$$p_{out} = p_{in} \cdot p_1 \cdot p_2 \cdot p_3 \cdot \text{etc} \quad (6.6)$$

where $p_j(j = 1,2,3, \ldots)$ represents the probabilities associated with each event outcome in the sequence. Expression (6.6) represents the usual assumption for event trees as employed in most conventional risk analysis work. It implies that all events are completely independent from one another.

Conditional aspects can be taken into account with the use of point estimates for the probability of one event occurring given that another has occurred. When such information is available, the generalization of (6.6) is:

$$p_{out} = p_{in} \cdot p_1 \cdot p_{2|1} \cdot p_{3|1,2} \cdot \text{etc} \qquad (6.7)$$

where $p_{i|j,k}$ represents the probability for the i-th event conditional on that for the previous events j and k. This describes the 'chain' of probability event along one event sequence, leading to each outcome of the event tree. Finally note that for each event in a sequence (or for any part of an event tree) the sum of the outcome probabilities must be equal to the sum of the probabilities coming in:

$$\sum_{i=1}^{n} p_{out,i} = \sum p_{in} \qquad (6.8)$$

6.4.2 Fault trees

Fault trees may be employed for overall, generalized system analysis (see Figures 3.2 and 6.2 for indicative examples). They may be used as input to an event tree, or a part of an event tree or on their own. Both applications are described in more detail in section 3.4.1. To see how the numerical analysis of a typical fault tree proceeds, consider Figure 6.2. Evidently, the lower events contribute to the top event and to the probability that a positive outcome occurs for the top event. Both AND and OR gates are shown. The combination of information contributing to the top event through an AND gate produces:

$$P_{top} = [E_a \cap E_b \cap E_c \cap \text{etc.}] \qquad (6.9)$$

When the events contributing to the top event are all independent, this becomes

$$P_{top} = p_a \cdot p_b \cdot p_c \cdot \text{etc.} \qquad (6.10)$$

while if they are fully dependent:

$$P_{top} = \max_i \{p_i\} \qquad (6.11)$$

Note that to deal with partial dependence between events it is necessary to use the conditional probability that one event will occur, given that the earlier one has occurred etc. This can be expressed, using the same notation as in Equation (6.7), by:

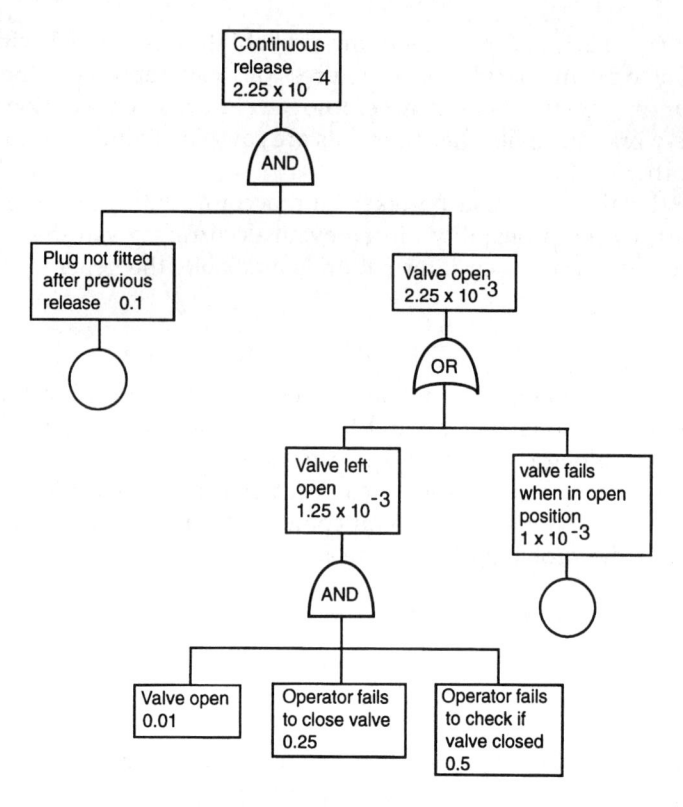

Figure 6.2 Fault tree.

```
                    Continuous
                    release
                    2.25 × 10⁻⁴
                        │
                       AND
              ┌─────────┴─────────┐
    Plug not fitted          Valve open
    after previous           2.25 × 10⁻³
    release  0.1                 │
                                OR
                        ┌────────┴────────┐
                   Valve left        valve fails
                   open              when in open
                   1.25 × 10⁻³       position
                       │             1 × 10⁻³
                      AND
           ┌───────────┼───────────┐
       Valve open  Operator fails  Operator fails
       0.01        to close valve  to check if
                   0.25            valve closed
                                   0.5
```

$$P_{top} = p_a \cdot p_{b|a} \cdot p_{c|b,a} \cdot \text{ etc.} \qquad (6.12)$$

If the contribution is through an OR gate then the corresponding statement is simply the union of the events:

$$P_{top} = P(E_a \cup E_b \cup E_c \cup \text{ etc.}) \qquad (6.13)$$

When the events in the set contributing to the top event are all independent, Equation (6.13) becomes simply the sum of the probabilities:

$$P_{top} = p_a + p_b + p_c + \text{ etc.} \qquad (6.14)$$

where now the probabilities are no longer conditional.

6.4.3 Integrated systems

When both types of trees are used, it is usual to have the output from a fault tree as input to an event tree (or part of an event tree). Figure 6.3 shows a simple example.

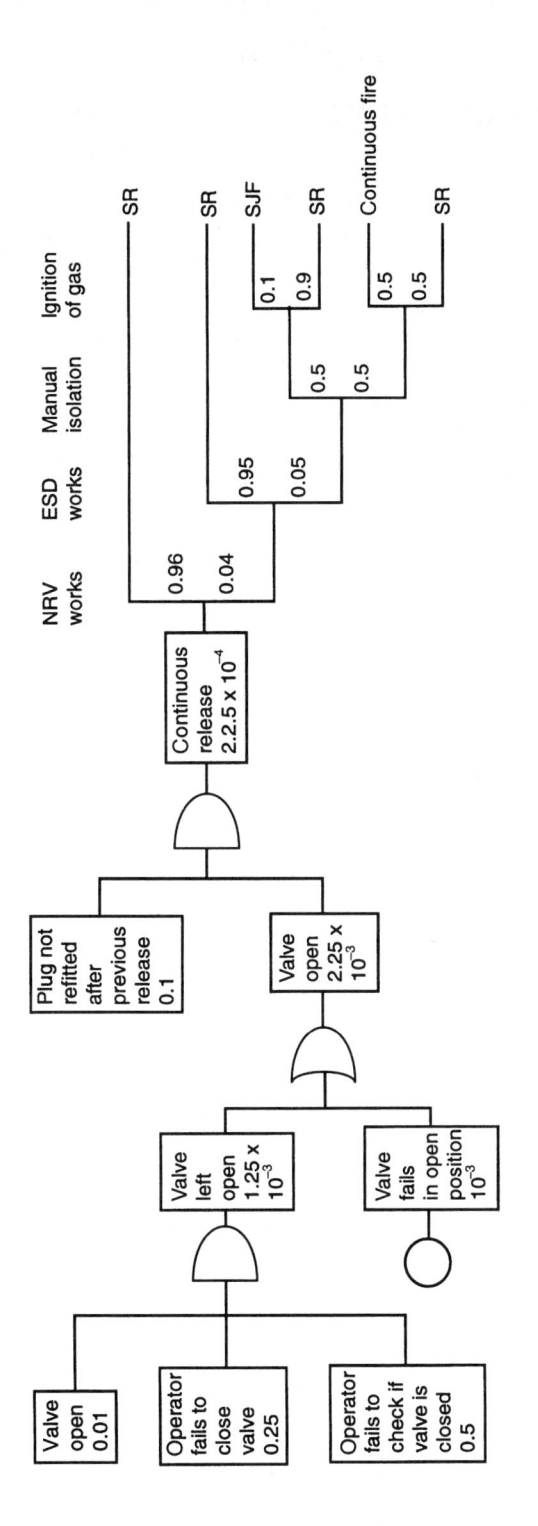

Figure 6.3 Combination of event and fault tree.

164 System evaluation

6.5 PROBABILISTIC RISK ANALYSIS (PRA)

6.5.1 Overview

In many cases the probability of occurrence of an outcome for an event cannot be described adequately by a point estimate value. An example is when there is a large amount of uncertainty about the probability of occurrence of the event outcome. In this case it is appropriate to retain the uncertainties associated with each event probability and to carry it forward to the next event. This has been termed 'error propagation' for Probabilistic Risk Analysis (PRA) in the nuclear industry (Apostolakis and Lee, 1977). It has been applied also in process plants (Kelly and Lees, 1986) and in other applications.

For a PRA it will be necessary to account for the uncertainties in the probability of occurrence of event outcomes. This means that appropriate information must be available. For example, a point estimate for the frequency of failure (however defined) clearly is not satisfactory for information such as given in Figure 6.4. This shows bounds on the frequency of failure to operate in a prescribed manner for valves of various types. Presumably, these bounds have been deduced from field and experimental observations, or they could also include subjective estimates. Evidently, the information represented by the bounds is of rather poor quality. The error bounds might suggest a rectangular probability density function as appropriate to describe the uncertainty (see Figure 6.5). However, it is unlikely that the characteristic of valve performance can be represented properly by such a function. If the data were re-analysed it might be possible to obtain a histogram of the data, and from this infer a more realistic probability distribution, such as the Gaussian (Normal) distribution shown in Figure 6.5. Note that the probability distribution inferred from the observational data now shows clearly that there is a level of uncertainty attached to the estimate for the event outcome, in

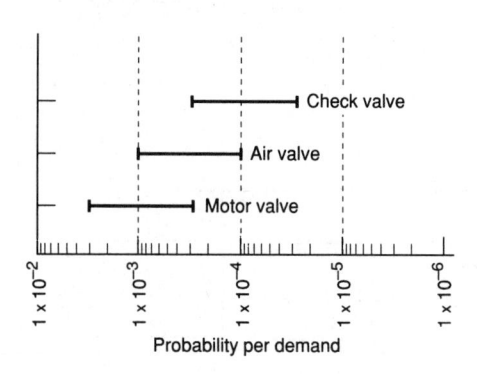

Figure 6.4 Valve operation data.

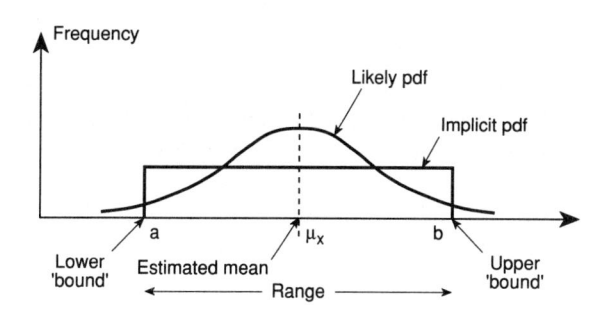

Figure 6.5 Implied and likely probability functions for data of Figure 6.4.

this case the frequency of valve failure (Melchers, 1993a). Thus event outcomes are treated as random variables (see also section 3.5.3).

Uncertainty about event outcome probabilities can be dealt with by employing: (1) 'second moment analysis' – a best estimate for the probability, together with a variance measure to describe uncertainty, or (2) full distributional analysis – a more advanced description involving higher statistical moments (i.e. when the frequency of failure itself can be described by a complete probability density function, such as shown in Figure 6.5). In either case, the analysis becomes more complex. These approaches will now be described.

6.5.2 Second moment analysis

In many cases information about the complete probability distribution of the uncertainty about event outcome probabilities is not available. Several different PRA techniques have been proposed to deal with this, including upper and lower limits, or quartiles (Mazumbar, 1982; Ragheb and Abdelhai, 1987) and various proposals for using means and variances (e.g. Jackson, 1982; Laviron, 1985; Kafka and Polke, 1986; Qin Zhang, 1989; Porn and Shen, 1992; Greig, 1993). The best of these is the approach described in terms of means and variances. In this case the outcome is described by the expectation (mean) μ of the probability of occurrence of the outcome and the associated variance σ^2, which is assumed to be available or can be estimated. In keeping with the work in structural reliability analysis it will be termed a 'second moment' analysis.

(a) Event trees

For the simple example event tree shown in Figure 6.1, the outcome for two events has a relationship directly corresponding to that of Equation (6.4), but now with uncertain inputs. This means the relationship becomes:

$$Y = X_1 \cdot X_2 \tag{6.15}$$

where Y and $X(i = 1, \ldots)$ are the random variables describing the input and event probabilities respectively. The expected outcome or, equivalently, the mean of the outcome μ_Y is given by

$$\mu_Y = E(Y) = E(X_1 X_2) = \mu_{X_1}\mu_{X_2} + \rho\, \sigma_{X_1}\sigma_{X_2} \tag{6.16}$$

and the uncertainty about this outcome is given by

$$\sigma_Y^2 = [(\mu_{X_1}\sigma_{X_2})^2 + (\mu_{X_2}\sigma_{X_1})^2 + (\sigma_{X_1}\sigma_{X_2})^2](1 + \rho)^2 \tag{6.17}$$

where ρ is the correlation coefficient between X_1 and X_2. If X_1 and X_2 are independent, as is often assumed to be the case in practice, then $\rho = 0$ and expression (6.17) simplifies to

$$V_Y^2 = V_{X_1}^2 + V_{X_2}^2 + V_{X_1}^2 V_{X_2}^2 \approx V_{X_1}^2 + V_{X_2}^{22} \tag{6.18}$$

where $V = \sigma/\mu$ = coefficient of variation and the right-hand approximation is reasonably close for V_{X_1} and V_{X_2} less than, say, 0.3.

The above shows how the uncertainty in the probability for each of the two event outcomes is carried through (propagated) to the two-component (sub-)system outcome, to give the uncertainty in the probability of the occurrence of the outcome. Again, results (6.16 and 6.17) may be used repeatedly as each further event is encountered in a sequence of events in a branch of an event tree branch. At each stage this process results in (1) best (i.e. mean) estimates for the probability of occurrence of each outcome in a sequence of events, and (2) estimates of the uncertainty associated with these probabilities.

(b) Fault trees

The treatment of uncertainty information for fault trees using 'second moment' information follows directly from the fundamental expressions for combination of event information. For AND gates, expressions (6.9 – 6.10) above apply directly. For OR gates, it is necessary to generalize (6.13) and (6.14) as follows. If X_1 and X_2 represent the uncertain probabilities associated with event outcomes for events 1 and 2, say, then the uncertain probability Y associated with the outcome of the two events contributing to the top event is given by:

$$Y = a_1 X_1 + a_2 X_2 \tag{6.19}$$

where, the a_i ($i = 1,2,3, \ldots$) are coefficients describing the way the sub-events contribute to the top event. The expected outcome is now given by

$$E(Y) = \mu_Y = a_1\mu_{X_1} + a_2\mu_{X_2} \tag{6.20}$$

and the variance by

$$\operatorname{var}(Y) = \sigma_Y^2 = a_1^2 \sigma_{X_1}^2 + a_2^2 \sigma_{X_2}^2 + \rho a_1 a_2 \sigma_{X_1}\sigma_{X_2} \tag{6.21}$$

where (X_1, X_2) is correlated through the correlation coefficient ρ. Evidently, if (X_1, X_2) are independent (as is often the case), the third part of expression (6.21) disappears since then $\rho = 0$.

(c) Example – 'Second Moment' event tree analysis

Consider the part of an event tree shown in Figure 6.6 (Melchers, 1993a). Let it be assumed that the input probability of occurrence is as shown, with the terminology (6.07 E-7, 1.0 E-7) representing the expected (mean) probability (6.07×10^{-7}) and its standard deviation (1.0×10^{-7}), respectively. The same notation will be used throughout this example. Let this probability be for the top ('yes') branch at the event 'RSD operable' (where RSD denotes 'Remote Shut Down'). The estimated probabilities and their associated standard deviations for event outcomes are shown in Figure 6.6. The overall outcome probabilities and the associated uncertainties (as represented by the standard deviations) can be obtained by working through the event tree using the above expressions for combinations of probabilities.

In Figure 6.6 it is evident that each of the last events considered has the upper branch with the outcome 'short jet fire' (SJF), a transient consequence. The total probability with which this consequence is estimated to arise is then given by the sum of several random variables:

$$\mu_{SJF} = (3.0E{-}7) + (2.97E{-}10) + (2.7E{-}12) = 3.0E{-}7$$

and

$$\sigma^2_{SJF} = (1.87E{-}7)^2 + (2.4E{-}10)^2 + (2.16E{-}12)^2$$

or

$$\sigma_{SJF} \approx 1.8E{-}7$$

Hence the contribution to overall risk from short jet fires is (3.0 E-7, 1.8 E-7).

(d) Outcomes from the analysis

The analysis procedures outlined above will result in estimates for probability of occurrence of particular system outcomes, each probability estimate being described in terms of a mean and a variance. This is rather different from conventional 'point estimate' procedures, and some consideration needs to be given to how this information might be interpreted.

Consider a simple case of one kind of system outcome, for which a prescribed criterion for acceptability exists (e.g. as set by regulatory authorities). The estimated mean for the expected probability of occurrence for this outcome is shown in Figure 6.7 as the mean μ together with a Normal distribution drawn to represent the variance σ^2 attached

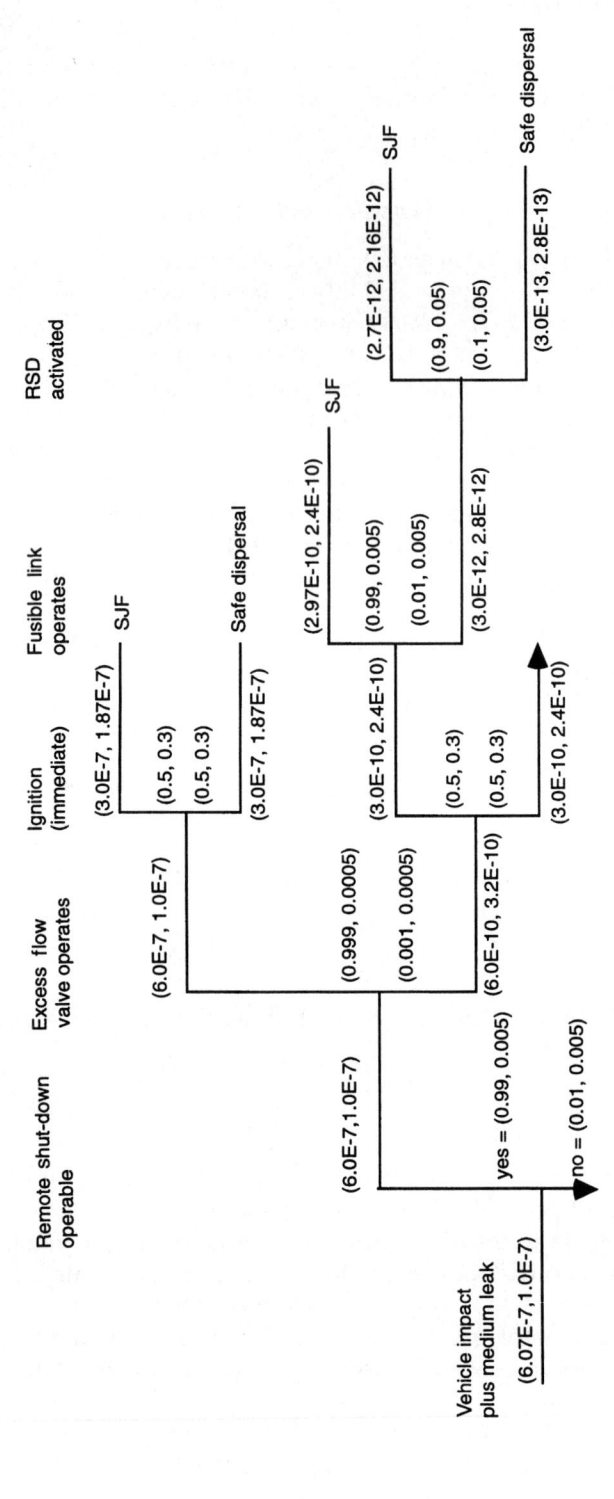

Remote shut-down operable	Excess flow valve operates	Ignition (immediate)	Fusible link operates	RSD activated	
				(2.7E-12, 2.16E-12)	SJF
			(2.97E-10, 2.4E-10)	(0.9, 0.05)	
			(0.99, 0.005)	(0.1, 0.05)	
		(3.0E-10, 2.4E-10)	(0.01, 0.005)	(3.0E-13, 2.8E-13)	Safe dispersal
		(0.5, 0.3)	(3.0E-12, 2.8E-12)		
	(6.0E-10, 3.2E-10)	(0.5, 0.3)			
	(0.001, 0.0005)		(3.0E-10, 2.4E-10)		
	(0.999, 0.0005)	(3.0E-7, 1.87E-7)	SJF		
	(6.0E-7, 1.0E-7)	(0.5, 0.3)			
		(0.5, 0.3)			
yes = (0.99, 0.005)		(3.0E-7, 1.87E-7)	Safe dispersal		
(6.0E-7, 1.0E-7)					
no = (0.01, 0.005)					

Vehicle impact plus medium leak
(6.07E-7, 1.0E-7)

Note: 1.5E-4 etc should be read as 1.5×10^{-4} etc.

Figure 6.6 Example: part of event tree.

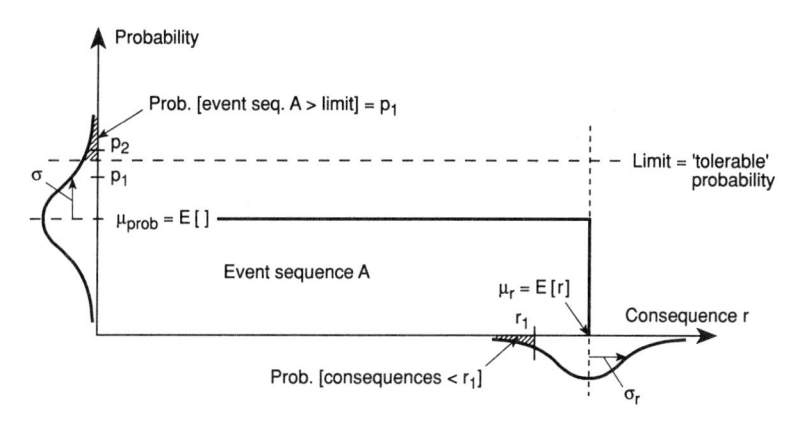

Figure 6.7 Probability of occurrence-consequence diagram.

to this estimate. It is clear that when the estimated probability exceeds the prescribed acceptance criterion, an unacceptable situation exists. The probability with which this is likely to occur is shown by the shaded zone. Note that Figure 6.7 illustrates neatly one of the fundamental deficiencies in data handling in conventional QRA – namely that 'conservative best' estimates for the various frequencies in event and fault trees provides an outcome probability of occurrence which is conservative to an unknown extent (Melchers, 1993a). Unfortunately, in conventional analyses there is no easy way to estimate the unknown uncertainty in the outcome probability estimate – reflection will show, however, that it may well be quite large.

The 'second-moment' type analysis used here uses the 'best' estimate (i.e. the mean or expected value) for each of the various probabilities of occurrence, with the result that the outcome probability is also a 'best' estimate (i.e. the estimate gives the expected value). Evidently, this is similar to the usual method of point estimates in QRA, except that mean values are used rather than conservative estimates. So in this sense there is no extra complication in the analysis. The major difference lies in the handling of uncertainty information. All the uncertainties in event outcomes (and represented by the respective variances) are eventually represented in the variance for the outcome probability. This is shown in Figure 6.7 in the 'spread' of the probability distribution on the vertical axis.

Figure 6.7 also shows how consequences ought to be handled. These are plotted on the horizontal axis. It is well-known that event outcomes, besides being uncertain in their probability of occurrence, are also uncertain in their magnitude or consequences. Thus the length of a flame (in the above example), the number of people killed or injured, the amount of damage which might be sustained, are all not easily predicted.

Probabilistic methods ought also to be used for these. As noted earlier, it is beyond the scope of this book to deal with the estimation of consequences, but it can be noted that they are best represented by an outcome probability density function, as sketched in Figure 6.7. Such uncertainty information can be very useful in estimates of damage. For example, for consequences in terms of flame length, the probability could be estimated of a particular facility at some known distance from the structure which would be affected. Further discussion of the representation of risk analysis outcomes is deferred to Chapter 7.

6.5.3 Full distributional analysis: Monte Carlo simulation

In many practical situations the data available for event outcomes will not be of sufficient quality to go much beyond the Second Moment approach given above. However, when sufficient data is available, high-quality probability models for the probability of event outcome (e.g. models such as an Exponential, or Gumbel distribution) can be constructed. A more complete analysis of outcome probabilities is then appropriate. The equations which need to be solved are still those for the combination of probabilities, such as Equations (6.5 and 6.9) for event and fault trees respectively. Some analytical techniques are available for this purpose, but these are seldom applied in practice.

A technique which has found much application in practical risk analysis problems is Monte Carlo simulation. It uses numerical simulation to solve Equations (6.5 and 6.9) so that the restrictions often demanded by analytical methods need not apply. Also, it can treat random variables having any type of probability distribution (including simplifications such as Second Moment representations) and it can deal with correlations between random variables, something not easily handled with 'point estimate' methods. Moreover, complex event trees and multiple event outcomes are easily handled.

The main disadvantage often noted for Monte Carlo simulation is its potentially high computational cost. Although this is still an issue for complex systems, it is increasingly less a problem due to the ready availability of powerful high-speed computers. In fact, it is doubtful whether with modern computers computation (CPU) times are actually a critical issue for the majority of systems likely to be encountered.

In its simplest form, a Monte Carlo simulation consists of many repetitions of a given sequence of calculations, each with randomly selected inputs. As applied in the present situation, this means repeating the analysis of event and fault tree with each analysis using different values of randomly generated event probabilities. The sequence of calculations in this case estimates whether the system fails (and how). The number of times such failure occurs can then be used to estimate the probability of failure of the system. Dependency between values can

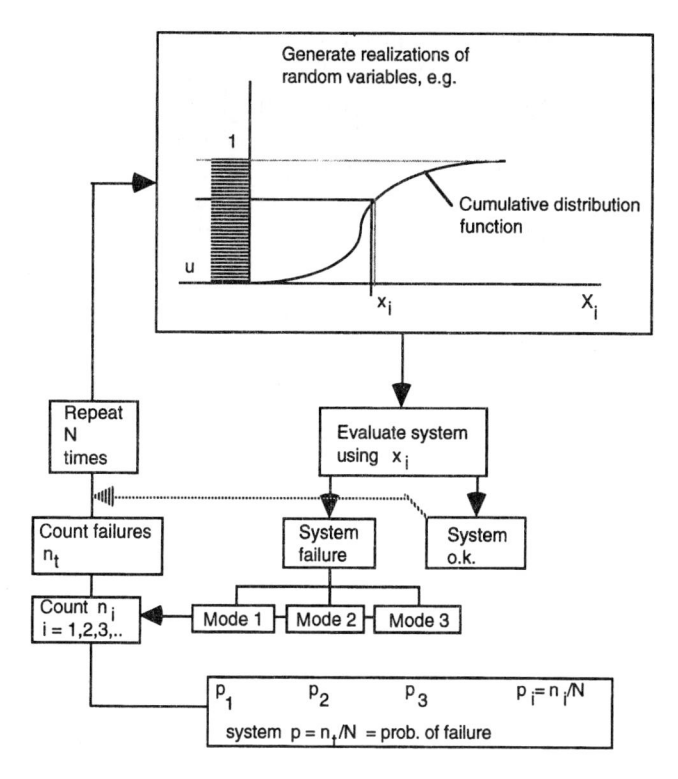

Figure 6.8 Schematic representation of Monte Carlo technique.

be accounted for using standard techniques for generation of dependent variables (e.g. Rubinstein, 1981).

The inputs for the calculations can be selected in a variety of ways, but, to allow for the uncertainty associated with the probability of event outcomes, the event outcome probabilities are considered as random variables with presumed known probability distributions. Random samples are then drawn from these distributions to give a deterministic set of occurrence probabilities. For each set of such sample values, the event tree outcome (failure or survival) is evaluated. The result is then taken as one sample outcome. When repeated many times, the sample outcomes from the event tree analysis can be analysed to obtain the histogram / probability density function for the output. This then forms the same type of result as shown in Figure 6.7 (above). The analysis process can be represented schematically as in Figure 6.8.

(a) Example – event tree analysis using Monte Carlo simulation

Some advantages of Monte Carlo simulation can be demonstrated from the following example. From the event-tree given in Figure 6.6, a Second

Moment analysis was used to calculate the risk from Short Jet Fires (SJF), see section 6.5.2(c). This analysis assumed that all event probabilities were Normally distributed. Clearly, the Normal distribution is unbounded ($-\infty$ to ∞), yet event probabilities actually should be bounded by 0 and 1. In such cases, the Normal distribution should be truncated (or censored) at 0 and 1. Analytical methods such as the Second Moment analysis cannot handle this requirement – Monte Carlo computer simulation analysis can. This process is now described.

Step 1. Event probability for 'vehicle impact plus medium leak' is randomly generated from Normal distribution (6.07 E-7, 1.0 E-7).

Step 2. If event probability is less than zero or greater than one then Step 1 is repeated.

Step 3. Repeat Steps 1 and 2 for other events ('RSD operable', 'EFV' etc.).

Step 4. Calculate risk of SJF by working through the expressions for combinations of event probabilities as described in section 6.4.1.

By replicating Steps 1 to 4 many times (i.e. by performing a large number of 'simulation runs') a simulation histogram of the probability of occurrence of SJF is obtained. From this appropriate statistical parameters can be inferred. Figure 6.9 shows the histogram resulting from 10 000 simulation runs (10–4 seconds per simulation run). The statistical parameters for this more realistic analysis were inferred to be (2.97E-7, 1.55E-7), values which are significantly different from those obtained from the Second Moment analysis, see Figure 6.9. Note that, given a similar number of simulation runs, the above Monte Carlo simulation analysis would yield results identical to those obtained from a Second

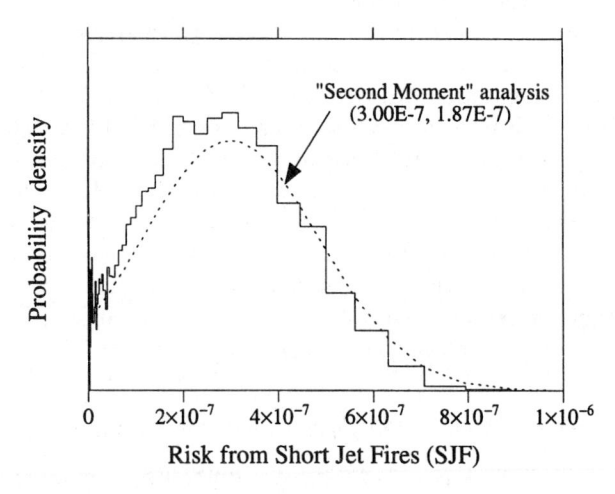

Figure 6.9 Simulation histogram for risk from Short Jet Fires.

Moment analysis if Step 2 had been omitted (i.e. event probabilities not bounded by 0 and 1).

(b) Refinement of Monte Carlo simulation

Evidently, the samples drawn for each random variable are likely to be clustered around the expected value (mean). As a result, the sample results for event tree outcomes are also likely to be clustered around the mean. Unfortunately, to produce good estimates for the probability density function describing the probability of occurrence of a system outcome (a random variable), samples away from the mean are preferred. In simple Monte Carlo simulation schemes the (unknown) uncertainty surrounding the output can only be estimated by taking more samples in the hope of getting a sufficient number well away from the mean. However, the problem can be overcome by adopting more refined methods of Monte Carlo simulation. These tend to reduce the number of samples required for a given level of accuracy in the event outcomes. They are known as 'variance reduction' techniques and include Latin Hyper-cube sampling and Importance Sampling (e.g. Rubinstein, 1981). A detailed discussion of some of these refinements will be deferred to section 6.8.7.

For efficient solution of risk analysis problems, it is increasingly the case that the analyst needs to give careful consideration to the structure of the problem, the degree of accuracy required in outcomes and the best way of modelling of the system. Generally it is advisable to simplify the problem as much as possible, consistent with the desired outcomes. However, the accuracy of the information available must be sufficient for a high degree of precision in modelling, considering also the amount of work involved. Also, attention needs to be given to whether the output is likely to be sufficiently meaningful, given the data available. In many cases a second moment type analysis might be sufficiently useful and accurate. An important exception is where the output probability distribution is likely to be significantly skewed.

6.5.4 Simplification of the system

It was already evident from the examples described in Chapter 3 that, for realistic systems, event and fault trees can become very large indeed. However, even for relatively simple systems the event and fault trees can be large, depending on the amount of detail, and hence the refinement in modelling that the analyst believes is required. The question then arises as to how to deal with such complex systems, given that there are constraints on the time and effort available for analysis and that the data available as input may not be of high quality. One approach is to do what is often done in practice: simplifying the event or fault trees in rather *ad hoc* ways through (1) deleting some parts of the trees or (2) shortening the trees by cutting back on analysis detail. This process of 'pruning' and 'lopping' of the trees is essentially one of 'truncation'. In some cases such a simple approach may be all that is warranted. However, ideally the truncation process should have a degree of rationality and should be such that it has minimal influence on the outcome of the analysis.

When the analysis is being done by hand, for an event tree, for example, it may be possible to work through some of the event probabilities at the same time and terminate a sequence of events once the probability of occurrence falls below a given criterion. For example, if system outcomes with a probability of occurrence less than, say, 10^{-6} are of little interest, a sequence of events with outcome expected frequency of, say, 10^{-8} might be terminated. Such an approach is often appropriate and is widely used.

For situations where (very large) event trees are automatically generated from a given set of events and known information about their connectivity, systematic techniques for truncation have been proposed. The conceptual basis for these various techniques is rather similar (e.g. Melchers and Tang, 1984; Thoft-Christensen and Murotsu, 1986).

6.6 SYSTEM DEPENDENCY

As already noted in Chapter 3, failure of multiple components due to system dependency effects can contribute significantly to the probability of failure of systems with redundancy or other forms of dependency. There are essentially two approaches for dealing with dependency: (1) explicit modelling and (2) implicit modelling.

6.6.1 Explicit methods

Explicit modelling of dependencies uses appropriate and careful event- and fault-tree development to ensure that functional dependencies are shown within the event or fault tree. In this way the event or fault tree deals directly with the dependencies. Evidently, this approach is best for obvious sources of dependence, such as might be caused by external events or agents, such as applied loads. Dependency between components or parts of the system can be presented in a dependency matrix, such as shown in Figure 6.10.

In general, for fault trees the modelling of dependencies is more difficult than it is for event trees. For this reason, various implicit methods have been proposed. These are discussed in the following section. In practice, there appears to be no clear distinction between when explicit or when implicit methods might be used. As a result,

	Probability of failure of item:			
Conditional on failure of:	i = 1	2	3	4
1	1	p(2|1)	p(3|1)	p(4|1)
2	p(1|2)	1	p(3|2)	p(4|2)
3	p(1|3)	p(2|3)	1	p(4|3)
4	p(1|4)	p(2|4)	p(3|4)	1

Note: p(i|j) denotes the probability of failure of 'i' conditional on 'j' having failed.

Figure 6.10 Dependency matrix (schematic).

there are variations in interpretation by system risk analysts and a degree of subjectivity remains. For example, they might well produce different lists of multiple failure events to be considered implicitly in an analysis.

6.6.2 Implicit methods

The implicit (or parametric) methods are useful to 'mop up' so-called 'common cause' failures which are not able to be identified during the analysis (so that the models have a 'catch-all' role). Interestingly, there appears to be some evidence that such common cause failures tend to have the same root causes as independent failures.

Implicit methods have been the subject of much discussion, misunderstanding and some controversy. According to Parry (1989), analysts make a deliberate choice not to represent explicitly each possibility (of different failure behaviour types) but instead to model their 'integrated' effect. The primary purpose of these models is to provide a framework for evaluation of the data. They should not be interpreted as theoretical models of the physics of failure, or as characterizations of the causal processes that result in either system or sequences of component failures. They were developed originally under conditions of scarcity of data – to a large extent this is still the case.

There have been objections to the implicit approach. The criticism is that in attempting to allow for dependency they develop point (or single-point value) estimates too early in the analysis. It would be preferable instead for the data uncertainty or variability to be kept as long as possible in the analysis. According to Doerre (1989), premature deterministic approximation of population variability violates a basic rule, which states that 'averaging is the last step in a stochastic calculation'. This view agrees with that expressed in the section above when dealing with error propagation in event and fault trees. Nevertheless, the implicit approach is often the only feasible approach in practical applications.

6.6.3 Implicit methods: system reliability cut-off method

In this approach, a 'ball-park' range of figures for the overall system reliability is estimated first. This may be done using experience for systems with similar basic design features, similar system redundancy and similar system diversity. The estimated range is then narrowed by considering the quality of the specific features of the system which act as defences against dependent failures. Such system defences are well recognized as important in overall system reliability (Watson and Johnston, 1987). They include:

1. design control, design review;
2. construction control, operational control;
3. reliability monitoring;
4. functional diversity, equipment diversity;
5. operational interfaces, protection and segregation;
6. redundancy and voting;
7. proven design and standardization, derating and simplicity;
8. construction control (standards, inspection, testing and commissioning); and
9. operational procedures (maintenance, proof testing and operation).

It is clear that this is a very subjective approach, but it may well be a useful starting point.

6.6.4 Implicit methods: common cause failure methods

Somewhat more detailed, but still 'short-cut', techniques also have been developed. In essence, these are based on the use of parametric models, with artificial events inserted in the fault tree, see Figure 6.11 (Fleming et al., 1986). It should be clear that AND gates in fault trees are of particular interest.

The parametric models used to describe the characteristics of the artificial events range from simple ones, such as the Beta factor method, to more complicated multi-parameter models such as the binomial failure rate (BFR) model, the multinomial failure rate (MFR) model, basic parameter (BP) model and the multiple Greek letter (MGL) model. A useful overview of the various models has been given by Watson and Johnston (1987). The topic continues to draw interest (e.g. Ansell and Walls, 1995).

Each model is based on a different set of assumptions. Thus each may require a different interpretation and each may require the available data to be used in a different way. Despite this, there is some evidence to indicate that with consistent use of data, models of similar complexity generate consistent results. The main source of lack of reproducibility and consistency appears to be dominated by the selection of parameters for the models. To some extent this is explained by the observation that

System dependency

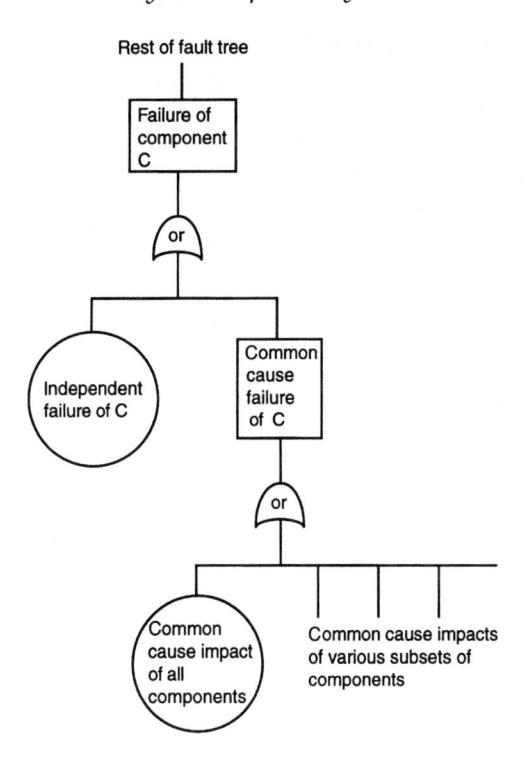

Figure 6.11 Common cause fault subtree for component A.

the models are essentially reformulations of one another. The data on which the models are based is critical, since the database for common cause failures is small, irrespective of the system under consideration. Moreover, existing data bases appear to be weak on situational and environmental characteristics. For this reason, analysts are forced to place considerable reliance on judgement, irrespective of the model used. Despite this, computer codes have been developed (Mosleh, 1991). Somewhat more detailed information about the more common methods is given below.

(a) Conventional Beta factor method

This is a simple approach, which tends to be used for dependency between identical components. The actual influencing factors are not modelled; instead the analyst attempts to relate the dependent failure contribution to a significant design feature of the system, typically a single component failure rate. It assumes, in essence, that if the component failure rate is reduced, the contribution of dependent failures also reduces. This assumption is not necessarily correct, since a decreased, or improved independent failure rate may not be accompanied by an increased robustness against external hazards and functional dependencies (such as common support/services).

The formulation for the Beta factor method assumes that the total failure rate λ for each redundant unit of the system can be decomposed into two additive parts – an independent and a dependent failure rate contribution, such that $\lambda = \lambda_i + \lambda_c$ respectively. Beta is then defined as:

$$\beta = \frac{\lambda_c}{\lambda_i + \lambda_c} = \frac{\lambda_c}{\lambda}$$

and is, in effect, the conditional probability of system failure given a single unit failure. If the components are identical (as is often the case), the system failure probability given by (6.9) would be simply $(\lambda T)^2$ where T is the (given) time interval of interest.

The dependent failure calculation would then be $\lambda_0 T = \beta \lambda T$, i.e. beta times the single component failure rate, and the overall system failure rate would be given by

$$(\lambda T)^2 + \beta \lambda T$$

For typical values of $\lambda = 10^{-6}$, $T = 10^3$ and $\beta = 0.1$ it is clear that the dependent component dominates the outcome. This is a typical result, and it emphasizes the importance of common cause failures.

(b) Partial Beta factor method

This method uses a greater degree of understanding about system-specific features. This allows a certain amount of structure to be added to the selection of the Beta factors. It starts with the assumption that there is a practical limit on Beta of $\beta = 2 \times 10^{-2}$ for identically redundant sub-systems (10^{-3} for diverse sub-systems).

The method proceeds by making an assessment of each of the various 'system defences' and giving a value to the associated partial Beta factor β_i (Watson and Johnston, 1987). The product of all such factors

$$\beta = \prod_i \beta_i$$

then provides the overall Beta factor, expected to lie in the range $10^{-3} - 1$. Evidently, the method blends the assessor's judgement with empirical data. Table 6.2 gives some typical values for partial beta factors.

(c) Multiple Greek letter method

The Multiple Greek letter method is an extension of the Beta factor method which allows consideration of different levels of redundancy (rather than only redundancy between two levels of components). For

Table 6.2 Typical partial Beta factors

Defences	Reference value	Assessed value
Design control	0.6	
Design review	0.8	
Functional diversity	0.2	
Equipment diversity	0.25	
Operational interfaces	0.8	
Protection and segregation	0.8	
Redundancy and voting	0.9	
Proven design and standardization	0.9	
Derating and simplicity	0.9	
Construction control	0.8	
Testing and commissioning	0.7	
Inspection	0.9	
Construction standards	0.9	
Operational control	0.6	
Reliability monitoring	0.8	
Maintenance	0.7	
Proof test	0.7	
Operations	0.8	
$\prod \beta_i$		

example, for a 4-component redundant sub-system, a new set of factors can be defined as follows:

β the conditional probability that a cause of component failure will be shared by one or more components,

γ the conditional probability that a cause of component failure shared by one or more components, will be shared also by two or more components,

δ the conditional probability that a cause of component failure shared by two or more components will be shared by all components.

These factors are then mutiplied as appropriate. Since typical values for β, γ and δ might be 0.1, 0.76 and 0.82, a large difference between this method and the simple Beta method is not likely. Other mean values for the Greek letter parameters may be obtained in the literature (e.g. Fleming et al., 1986).

6.7 EFFECT OF TIME

The techniques for risk analysis described above have all dealt with a very specific situation: namely, given a system, with expected behaviour of various components (and perhaps some measure of the uncertainty surrounding such behaviour), what is the estimated probability of the system failing to perform (to some defined criteria)? This approach implies that the assessment is done with information relevant to some particular point in time. However, the risk to a system or facility may change with time either due to its general deterioration, under processes such as fatigue and corrosion, and/or due to the changing nature of the loadings (environmental or man-induced). Some account for these changes may be made by tailoring the analysis to the particular point in time. This could be the present, using existing and current information, or for some future point in time, using extrapolated data and models. Additionally, in the analysis discussed so far, no attention was given to the question of the time which would be expected to elapse before failure might occur in the future. This requires, of course, estimation of the rates of failure for those components which are likely to be affected by the effect of time. This adds extra complexity and uncertainty to the analysis.

Time is also important in the representation of loads or demands such as wind loads, which for many purposes are better treated as stochastic processes rather than random variables. Generally it is assumed that such processes do not change their probabilistic nature with time; they are then termed 'stationary' processes (see also section 4.2.5). Procedures to deal with stochastic processes are available, but will not be treated in detail herein.

6.7.1 Effect of time on failure rates

As noted in section 4.2.1, it has been observed that the behaviour of system components, and indeed some total systems, can be described by a so-called 'bath-tub' curve (see Figure 4.1). The bathtub curve consists of three distinct zones:

1. 'burn-in' phase,
2. useful life phase and
3. 'wear-out' phase.

What is clear is that the failure rate is not at all constant, except perhaps approximately for the section in the middle of the component life. Yet the assumption of a constant failure rate underlies much conventional reliability theory for mechanical components. The reason for this has to do with ease of mathematics rather than sound probabilistic modelling, as will now be discussed.

The failure rate above is more properly termed the 'hazard function' $h_T(t)$ (also known as the 'age-specific failure rate' or the 'conditional failure rate'), defined as the probability that the system will fail at time T, given that the system has not failed prior to time t:

$$h_T(t) = \frac{f_T(t)}{1 - F_T(t)} \quad (6.22)$$

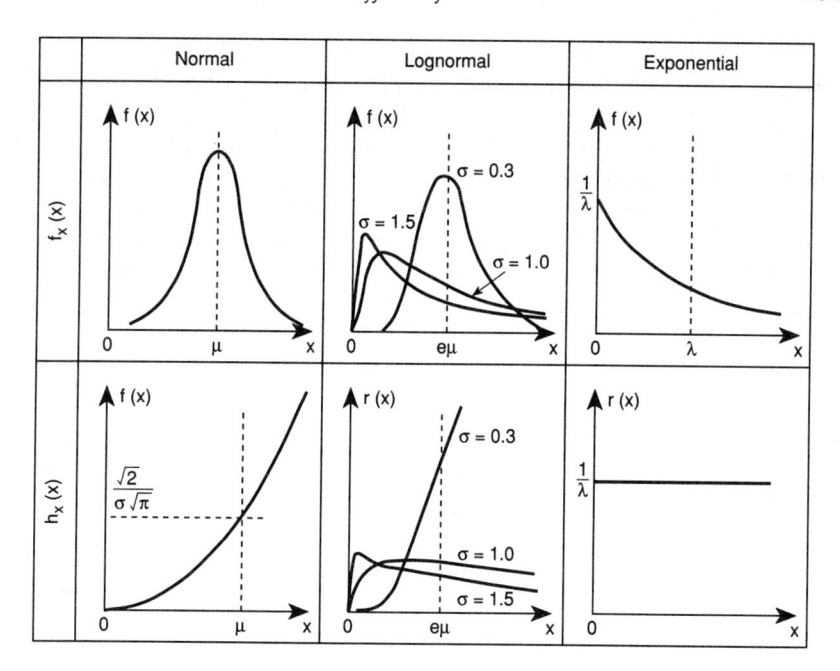

Figure 6.12 Typical hazard functions.

where $f_T(t)$ is the probability density function and $F_T(t)$ is the cumulative distribution function for the time to failure t. Hence if the probability density function and the cumulative distribution function are known, the hazard function can be obtained immediately. As already noted in Chapters 4 and 5, the required probabilistic descriptions can be obtained or estimated, for simple components, from observations of component life. For more complex components or for sub-systems, the techniques to be described in section 6.7.3 may be used.

Various forms of hazard function and their relation to the probability density function for time to failure t are shown in Figure 6.12. It is clear that only the exponential distribution corresponds to a constant hazard rate (i.e. it is constant with time).

6.7.2 Time to first failure for components

In system evaluation, the time which is expected to elapse before the occurrence of the first mode of system failure is of obvious interest. This could result, for example, from the first occurrence of failure of a critical component or from the first occurrence of a sufficiently high environmental loading. Further, in multi-component systems the failure of a significant number of components may be necessary to result in system failure. In this case the time to first failure of the system as a whole is of interest. Hence a relationship is needed between the time to first occurrence of component failure and the time to first occurrence of system failure.

The 'mean time to failure' (MTTF) or the 'mean life' for an individual component is obtained by deduction from data based on long-term observation of many (and if possible) identical components. If the data gives the life-times $t_1, t_2, \ldots t_n$ of n nominally identical components, then the MTTF is given by (e.g. Green and Bourne, 1972; Shooman, 1968; Smith, 1985):

$$MTTF = \frac{1}{n} \sum_{i=1}^{n} t_i \quad (6.23)$$

For components which are replaced when failure occurs (i.e. the component is replaced by a nominally identical component), the MTTF can be estimated also from the total operating time T_{op} and the number of failures k in that time:

$$MTTF = T_{op} / k \quad (6.23a)$$

In this form the MTTF is the same as the 'return period' T_R used extensively in civil and structural engineering applications. It is the time expected to elapse before the occurrence (or re-occurrence) of a defined event. This may be failure of a component, a sub-system, the total system, or the occurrence of some other defined event, for example, the occurrence of a wind speed greater than a given magnitude. Typically the return period is measured in years. In this case the reciprocal of the return period will be the (conditional) probability p_t of failure (or event occurrence) per year, which shows that:

$$T_R = \frac{1}{p_t} = \frac{1}{h_T} = \frac{1}{\lambda} \quad (6.24)$$

where h_T is the 'hazard rate' defined in (6.22) above. The term λ is introduced here as it is the symbol often used to denote the (age specific) 'failure rate', also referred to above as being equivalent to the hazard rate for the specific situation here.

The electrical and mechanical reliability literature also refers to terms such as the mean time between failure (MTBF), down time, mean time to repair (MTTR), etc. (e.g. Smith, 1985). In principle the meaning of these terms ought to be clear. However, some confusion between MTTF and MTBF does arise. The latter applies strictly only to components (sub-systems) which are repaired or replaced (i.e. a 'renewal' situation). MTTF and MTBF are equivalent only in constant hazard rate situations. This is usually the case for naturally occurring environmental loads, as these are modelled well enough as stationary continuous processes. For mechanical and electrical components MTTF and MTBF generally are not equivalent, as reflection about the meaning of the bath-tub curve will show. The mean time between failures (MTBF), down time, mean time to repair (MTTR) generally have little direct relevance to the risk assessment of a high reliability system.

6.7.3 Time to failure and probability of failure for systems

The probability of failure and the time to first failure for a system (or sub-system) can be obtained from the hazard rate $h_T(t)$ for the components which make up the system, as will now be described.

The probability of system failure over a total time period $0 - t$ will depend on the probability that failure does not occur in the first elemental period $(t_1) - (t_1 + dt)$, followed by no failure in the second elemental time period, given there was no failure in the first, and so on. For any elemental time period, the hazard rate h_i for that period expresses the probability of failure given that failure has not occurred earlier, hence the probability of no failure at all in the period $0 - t$ is given by:

$$p(0) = (1 - h_1) \cdot (1 - h_2) \cdot (1 - h_3) \ldots \quad (6.25)$$

The probability $p(t)$ of the system failing in the time period $0 - t$ is given by $p(t) = 1 - p(0)$. Also, noting that $(-p)^n \approx -\exp(np)$ and letting all the h_i be equal to λ it follows reasonably readily that

$$p(t) = F_T(t) = 1 - \exp\left[-\int_0^t \lambda(\tau)d\tau\right] \quad (6.26)$$

which reduces, for constant hazard rate, to:

$$p(t) = F_T(t) = 1 - e^{-\lambda t} \quad (6.27)$$

or, for a very low probability of failure of the system:

$$p(t) \approx \lambda t \quad or \quad \approx \int_0^t \lambda(\tau)d\tau \approx \sum_{i=1}^n \lambda_i t_i \quad (6.28)$$

where λ_i is the constant hazard rate in the i-th time period t_i. From this expression it can be seen that the hazard rates for each elemental time period can be added, provided they are approximately constant during each such time period.

In risk analyses for electrical and mechanical components it is common to assume that the components have lifetimes which can be described by the exponential function, which means that they all have a constant hazard rate function (see Figure 6.13). It follows from Equation (6.28) that the use of constant hazard rate functions considerably simplifies system risk calculations. This will be illustrated for two particular idealized system configurations.

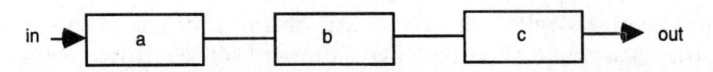

Figure 6.13 Two component series system.

For a series system, such as shown in Figure 6.13 with the components a, b, c etc., the system will fail when any one of the components fails. This can be expressed by:

$$1 - p(t) = [1 - p_a(t)][1 - p_b(t)]. \ldots \quad (6.29)$$

If all components have constant hazard rates (6.27) can be applied directly, so that:

$$e^{-\lambda t} = e^{-\lambda_a t} \cdot e^{-\lambda_b t} \cdot etc. \quad (6.30)$$

The (series) system hazard rate is then given by $\lambda = \lambda_a + \lambda_b + \lambda_c + etc.$ and the probability of failure of the system for this particular component failure sequence is given by:

$$p(t) = 1 - \exp[-(\lambda_a + \lambda_b + \lambda_c + etc.)] \quad (6.31)$$

This shows the advantage of assuming constant hazard rates for systems composed of components connected or acting in series. For complex systems usually there will be a number of possible series system failure combinations. The total probability for all such combinations forms a parallel system and is obtained as follows.

In an idealized parallel system it is necessary for a sufficient number of components to fail before the system will fail. Consider a simple case – one in which both components of a two component system must fail to cause system failure (see Figure 6.14). The more general case follows readily. The system failure probability is now given by Equation (6.5) which becomes, for independent components:

$$p(t) = p_a(t) \cdot p_b(t) \quad (6.32)$$

or

$$p(t) = [1 - \exp(-\lambda_a t)][1 - \exp(-\lambda_b t)] \quad (6.33)$$

In general there will be a number of possible combinations of component failure which will cause system failure. For constant hazard rates

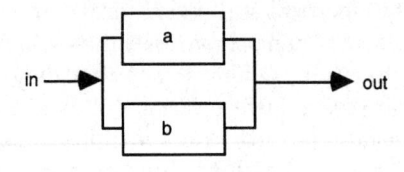

Figure 6.14 Two component parallel system.

and low probabilities of failure, as assumed here, the probabilities of failure for each parallel failure mode can be added according to Equation (6.28).

6.8 RELIABILITY OF LOAD-RESISTANCE SUB-SYSTEMS

6.8.1 Basic concepts

The systems considered in the previous section comprised components and other system elements for each of which the probability of a particular outcome of an event could be estimated either directly from data based on past observations or indirectly from subjective estimates (Bayesian estimates). Moreover, provided the system was analysed in sufficient detail, dependencies between events could be accounted for in a relatively straightforward manner.

Attention is now turned to systems and sub-systems for which individual event outcome probabilities cannot be estimated directly from observed data but must be calculated, using probabilistic – or stochastic – information. These sub-systems or systems are represented as 'load-resistance' (or 'demand-capacity') system elements. They are encountered in structural, mechanical, electrical and civil engineering systems. Note that the probability calculated for a 'load-resistance' element or sub-system is incorporated into a QRA or PRA if the element is part of a larger system.

To illustrate the situation of interest, consider a tower subject to wind forces. The probability of the tower toppling under the action of the continuously varying but probabilistically described wind force will depend on the strength of the wind and also on the uncertain strength of the tower. The strength, in turn, might depend on the individual strength of several members in the framework forming the tower structure. These member strengths may be highly correlated, since they could have been fabricated from the same billet of steel. The loss of strength of any one member does not necessarily lead to collapse of the tower. Whether the tower fails will depend on the structural configuration and the member being considered. Nor does it follow that the maximum historically recorded wind force is that which will cause collapse – some lesser wind force might be sufficient.

The illustration can be extended. Let now the tower be part of a series of towers making up an electricity supply system. Obviously, failure of any one tower could lead to fracture of the transmission wires and hence failure of supply. Alternately, if there is redundancy in the system (the usual case), some tower failures might be tolerable. In either case, it is likely that there will be a degree of correlation between the behaviour of the various towers, for the following reasons:

1. similar wind forces are likely to be experienced by at least some of the towers at much the same time (due to the wind field effect),
2. the transmission wires supported by the towers will tend to propagate failure along the system should one tower fail, and
3. the towers might well have been constructed from similar batches of steel, so that if one is under-strength, they are all likely to be under-strength.

It should be clear from this example that the basic principles of event and fault trees will also hold for the analysis of overall systems of this type (i.e. the electricity supply system above). The dependence structure in the overall system is obviously an important aspect to be considered. Also, the analysis required for some of the main sub-systems in the system can be quite complex – for example, the determination of the probability of failure of a single tower under wind and other loads. This is a 'load-resistance' element that constitutes a sub-system within the overall system.

The main focus of the present section will be on how to deal with the probabilistic analysis of sub-systems subject to time-dependent loading. This may be due to natural phenomena (such as wind, waves, snow, earthquake etc.) or it may be due to man-made effects (such as crowd loading, vehicular loading, industrial processes etc.). A discussion of the types of loads and resistances and the manner in which they can be modelled has been given already in section 4.4. Such information will now be used to estimate the probability of sub-system failure.

6.8.2 Simple formulation

The basic ideas for dealing with (sub-)systems subject to stochastic processes or described by random variables in a risk analysis problem were introduced in section 4.4. They will now be discussed in more detail. Consider a system having a capacity (or resistance) R. The system is likely to fail when the demand Q (e.g. an applied load) exceeds R, i.e. if $Q > R$. If the demand Q is now to be considered as a stochastic process the actual magnitude will be a random variable at any point in time, with the magnitude described by some probability density function. A typical realization of the process is shown in Figure 6.15. Similarly, the capacity R can be modelled as a random variable, stationary in time or slowly varying (as in gradually deteriorating systems,) such that $R = R(t)$. It is also possible to have R as a stochastic process, although this is relatively unusual.

The matter now of interest is the probability of the demand or loading exceeding the capacity or strength. For realistic systems it is evident that, given sufficient time, this will be a certain failure event (as a result of deterioration, for example). Hence, the time which will be expected to

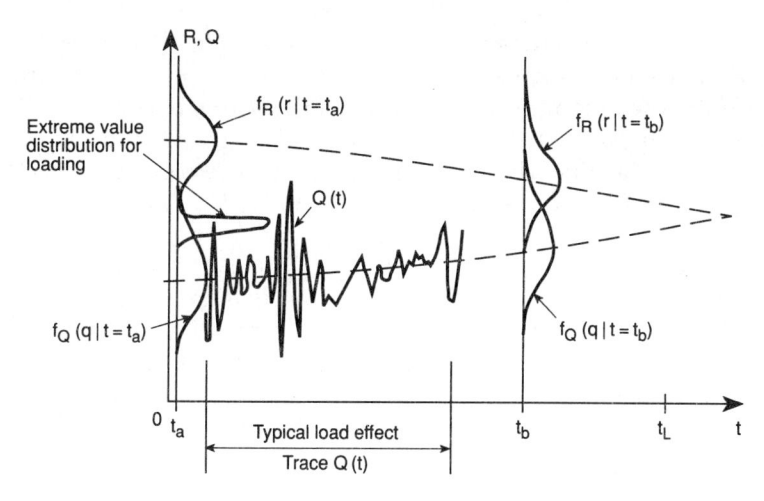

Figure 6.15 Time-dependent reliability.

elapse before the first occurrence of the failure condition $Q > R$ is of obvious interest. As noted in section 6.7.2, in conventional reliability work this time period is referred to as the mean time to failure (MTTF), defined in Equation (6.23). When dealing with stochastic processes this time is referred to as the time to 'first exceedence' and is calculated either from an 'up-crossing' analysis (when one stochastic process is involved), or an 'out-crossing' analysis (when more than one stochastic process is involved).

In general the problem posed above is rather difficult to solve. It will be useful for most cases to employ important simplifications to the problem formulation. These will allow reasonably simple methods for estimating (event) failure probabilities to be developed (Madsen et al., 1986; Melchers, 1987, 1993b).

Let it now be assumed that the (sub-)system of interest will fail under the maximum demand or load which could occur during the 'lifetime' t_L (however defined) of the sub-system. Rather than using the complete stochastic process, only the probability distribution describing the 'maximum' or the 'extreme value' of the process will be employed. Extreme value distributions, such as the well-known Gumbel distribution, can be used to model applied loading or demand. In order to construct the distributions, the parameters of the extreme value distribution are obtained from discrete observations of the demand or load process over a period of time. This usually means that the maximum demand or load observed each year is the recorded value used for probability distribution estimation. This is a special case of a stationary applied demand or load process, that is, one in which the distribution parameters such as the mean, variance etc. do not change with time.

The instantaneous probability distribution for the process is then as shown at the left of Figure 6.15. Also shown is a schematic extreme value distribution for the maximum in the demand/load process.

The probability $p(t)$ that the maximum demand or load Q_{max}, when applied for the first time to the system, will exceed its capacity or strength R can be expressed now as:

$$p(t) = P[R(t) \leq Q_{max}] = P[R(t) - Q_{max} \leq 0] \tag{6.34}$$

Let now the capacity or strength R be taken as constant with time (or simply consider an appropriately reduced capacity at some time point of interest). Then the uncertainty about R can be represented by a random variable quantity with cumulative density function $F_R(r)$ so that expression (6.34) becomes the well-known 'convolution' integral (see also Equation (4.5)):

$$p = P[R - Q \leq 0] = \int_{-\infty}^{+\infty} F_R(x).f_Q(x)dx \tag{6.35}$$

which is the (unconditional) probability of failure under the application of the maximum load Q_{max} (a random variable with probability density function $f_Q(q)$) acting on the system when the system has a strength denoted by the random variable R. Note that this formulation is 'time-independent', since time has been absorbed as a parameter in the way the loading has been defined; for Q based on annual maxima, the probability determined by (6.35) will be the annual probability of failure. It follows that the probability of failure determined from (6.35) is a conditional result. It depends on the way the random variables R and Q have been defined and any conditionality imposed on their respective distributions.

For elementary cases the integration in (6.35) is not difficult. It can be performed by numerical integration and, in addition, charts and tables are available for the solution of (6.35) for a range of probability density functions (Melchers, 1987). For the very special case where both Q_{max} and R are described by the Normal distribution, a very simple analytical result can be obtained. The Normal distribution may be described by its mean μ and its standard deviation σ. Also, a new random variable $Z = R - Q$ is defined, with, obviously, $Z < 0$ denoting failure. It follows that $\mu_Z = \mu_R - \mu_Q$ and that $\sigma^2_Z = \sigma^2_R + \sigma^2_Q$. The probability that $R \leq Q_{max}$ then becomes:

$$p = P[R - Q_{max} \leq 0] = P[Z \leq 0] = \Phi\left(\frac{0 - \mu_Z}{\sigma_Z}\right) \tag{6.36}$$

where $\Phi()$ is the well-known standard Normal distribution function (zero mean, unit variance) extensively tabulated in statistics texts.

6.8.3 Generalized formulation

For many realistic problems the simplified formulation (6.35) is not sufficient. Usually several random variables will influence the capacity or resistance of the sub-system. Thus, for a strength problem, the resistance might be a function of material tensile strength, cross-sectional area, temperature etc. It follows that in general the resistance or capacity of the sub-system is a vector function of various parameters. Let these parameters be denoted by the vector **X**. It follows that $R(\mathbf{X})$ needs to be substituted for R in the above expressions.

In addition, there are likely to be several load processes acting on the system at the same time. For example, these might be wind, wave, temperature and pressure forces acting on an offshore structure. In this case, account will need to be taken of the (generally very low) probability of the simultaneous occurrence of peak loads. This is a rather complex issue and will not be addressed specifically herein. A relatively simple approach is to let the stochastic load processes be replaced by appropriate random variables. Thus one approach is to use the distribution for the maximum value (e.g. annual maxima, as was done above) for one of the loads and the distribution for the average load for all the others. All combinations of maximum and average loads should be used (the so-called 'Turkstra's rule' (Turkstra, 1970), see also section 4.4.4. The net result is that the loadings can now be treated as a vector of random variables, and let these be included in the vector **X**. (Note that it also possible for the parameters of the random variables (such as the loads) themselves to be random variables, describing uncertainty in their values.)

In the vector $\mathbf{X} = \{X_1,\ldots,X_n\}$ of random variables each component represents some resistance random variable or some loading random variable acting on the system. This vector will possess a joint probability density function $f_x(\mathbf{X})$. The probability of failure may now be described as:

$$p = \int_{G(X)\leq 0} f_x(\mathbf{x})\, d\mathbf{x} \qquad (6.37)$$

where the function $G(\mathbf{X}) \leq 0$ is the so-called 'limit state' function, which describes the region of unacceptable system behaviour (the failure region). It is a generalization of Z in expression (6.35) above. It is obtained from considering that combination of random variables which describes the failure state(s) for the event. In practice, the function $G(\mathbf{X}) \leq 0$ may be discontinuous and composed of several component functions, which together describe the 'acceptable' region, see Figure 6.16. In particular, it shows a 'series' system description of a set of limit state component functions, expressed mathematically as:

$$\bigcup_{i=1}^{m} G_i(\mathbf{x}) \leq 0 \qquad (6.38)$$

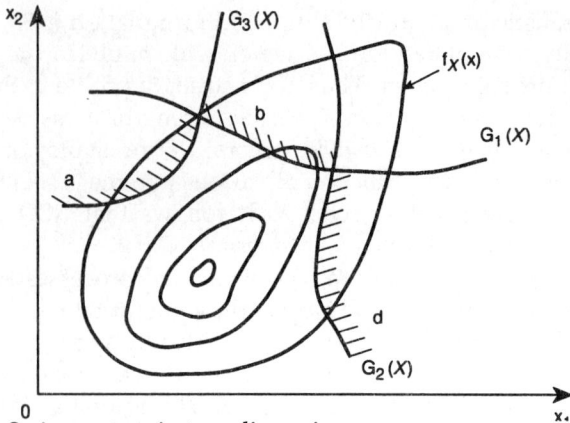

Figure 6.16 Series system in two dimensions.

while Figure 6.17 shows a set of limit state function for which the 'parallel' system description

$$\bigcap_{i=1}^{m} G_i(\mathbf{x}) \leq 0 \qquad (6.39)$$

is appropriate. Note that in each case the condition ≤ 0, by convention, describes the 'unacceptable' region.

Equations (6.37–6.39) add considerably to the complexity of the calculation of the failure probability. While numerical integration of (6.37) may be possible for a very small number (say less than 5) random variables, it quickly becomes extremely demanding of computer time, and generally is not considered to be a feasible approach. Fortunately, techniques to handle such problems have been developed. In the following only two broad classes of solutions will be described. These are the so-called First Order Second Moment methods (and developments thereof) and Monte Carlo simulation methods (Madsen *et al.*, 1986; Melchers, 1987).

6.8.4 First Order Second Moment (FOSM) method

(a) Basic case – linear limit state function

In the First Order Second Moment (FOSM) method the terminology 'second moment' refers to the description of all the random variables in terms only of their mean (the 'first moment') and their variance (the 'second moment'). The Normal distribution is adequately specified by these two 'moments' and for simplicity (and applications) one could read

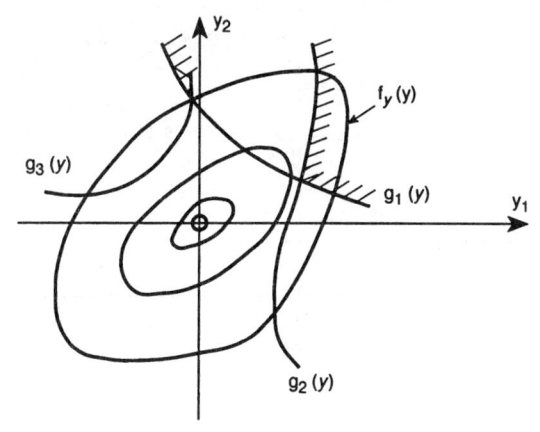

Figure 6.17 Parallel systems in two dimensions.

'second moment' methods simply as performing calculations only with Normal distributions. It would be misleading, however, to interpret the method only as valid for Normal distributions – its application merely implies that whatever the distribution attached to a random variable, only the first two moments are considered for calculation purposes.

The term 'first order' defines the expression for the failure condition as a linear function. In section 6.8.2 above, the failure condition was given by $Z = R - Q$ which is clearly linear. In general, however, particularly where there are a number of random variables involved, or where the 'limit state function' is a composite of several components, such as in (6.37) or (6.38), the function' is very likely to be nonlinear. In these cases, linearization of the limit state function can be achieved through a Taylor series expansion of the function about some appropriate point (see below).

For the case where the limit state function is indeed linear, use can be made of a very simple result. Let each of the random variables be transformed to the so-called 'Standard Normal' space U, such that each transformed variable is standard Normal, with zero mean and unit variance. The limit state function needs also to be transformed to the standard Normal space and is denoted $g(U) \leq 0$. Figure 6.18 shows the result in two dimensional U space. The limit state function(s) defines 'unacceptable' regions away from the origin, i.e. the acceptable region surrounds the origin. Obviously the limit state closest to the origin will be the critical one. It is also the limit state function which has the highest probability content associated with it – this is the 'volume' under the standard Normal shape within the 'unacceptable' region. Hence the integral (6.37) is a volume integral in two dimensions, as in Figure 6.19. In general, X (and hence U), is a multi-dimensional vector of

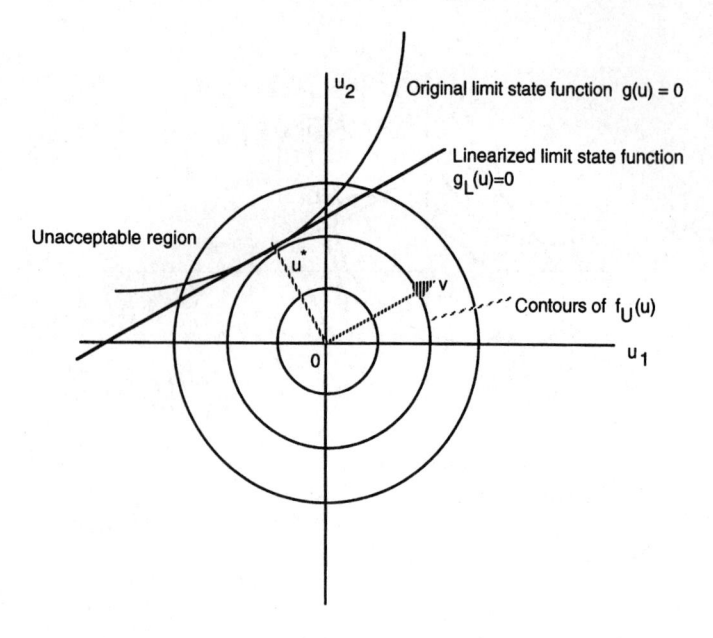

Figure 6.18 Linear and non-linear limit state functions in standard normal space.

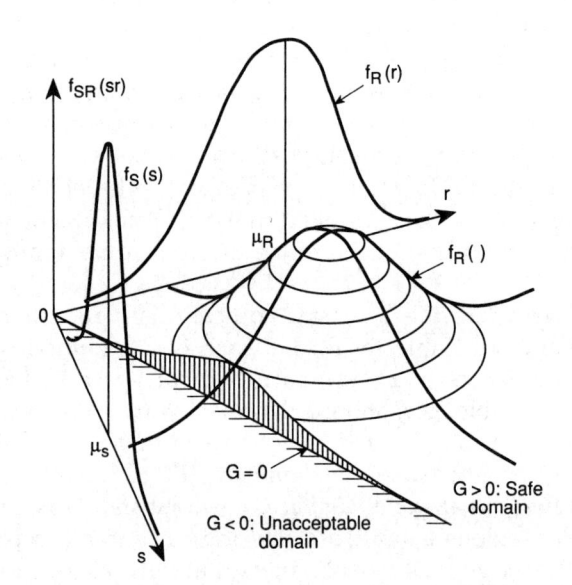

Figure 6.19 Region of integration for failure probability.

[Figure: Marginal distribution in standard normal space, showing g(y)=0, β, p_fN, and 0 on axis]

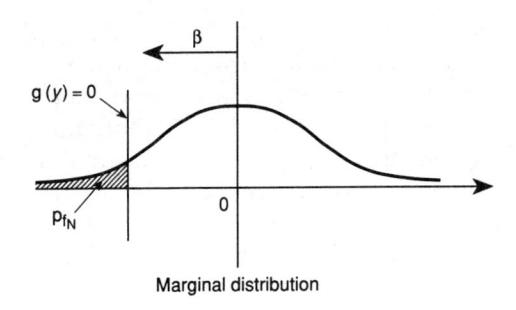

Figure 6.20 Marginal distribution in standard normal space.

random variables, so that the integral (6.37) is a multi-dimensional 'volume' integral.

The probability content in the 'unacceptable' region for the (single) linear limit state function in Figure 6.18 can be obtained directly as

$$p = \Phi(-\beta) \qquad (6.40)$$

where β represents the (shortest) distance from the origin to the linear limit state function. (In the structural reliability literature β is known as the 'safety index'. It can be used as a substitute for the failure probability.) Evidently, if $\mathbf{u}^*$ describes the coordinates of the point closest to the origin, then

$$\beta = \min\left(\sum_{i=1}^{n} u_i^2\right)^{1/2} = \min(\mathbf{u}^{*T}.\mathbf{u}^*)^{1/2} \qquad (6.41)$$

This result also follows from integrating at right angles to the limit state function in the standard Normal space. It can be interpreted as in Figure 6.20, obtained from integration in the v direction in Figure 6.18. The probability content corresponding to the 'unacceptable' region (the shaded zone) is obtainable from standard Normal tables. (Note, however, that the tables typically give values for the positive unshaded part of Figure 6.20.)

(b) Non-linear limit state functions

If the limit state function is not linear, the FOSM method can still be applied if the limit state function is linearized. First it is noted again that the greatest contribution to the probability of failure is made by those points closest to the origin in standard Normal space U. Also, the linearization should be such as to permit estimation of the probability content as best as possible. Intuitively, the best point about which to linearize (using a Taylor series expansion) is the point $\mathbf{u}^*$ closest to the origin and positioned on the (non-linear) limit state function. This point is sometimes known as the 'design' or 'checking' point. It could be obtained from expression (6.41) using mathematical programming. Alternatively, a trial and error procedure could be used, or iterative approximation procedures. Such systematic procedures are necessary in the multi-dimensional space required for realistic problems.

The concepts involved for a non-linear limit state function can be illustrated using a Lagrangian multiplier when the limit state function is differentiable. This will also bring out some of the technical features skipped over in the above discussion.

(c) Example

This example (after Melchers, 1987) is a simple case in which there are only two random variables, already transformed to the standard Normal space. Let the limit state function be $g(\mathbf{u}) = -\frac{4}{25}(u_1 - 1)^2 - u_2 + 4 = 0$. According to Equation (6.41) the distance to be minimized is given by $(u_1^2 + u_2^2)^{1/2}$ under the condition that the limit state function is satisfied. The objective function for minimization then becomes, with λ the Lagrangian multiplier,

$$\min(\Delta = (u_1^2 + u_2^2)^{1/2} + \lambda \left[-\frac{4}{25}(u_1 - 1)^2 - u_2 + 4 \right] \quad (6.42)$$

To obtain the minimum, the first derivatives with respect to (u_1, u_2, λ) must be set to zero:

$$\frac{\partial \Delta}{\partial u_1} = u_1(u_1^2 + u_2^2)^{-1/2} - \lambda \frac{8}{25}(u_1 - 1) = 0 \quad (6.43a)$$

$$\frac{\partial \Delta}{\partial u_2} = u_2(u_1^2 + u_2^2)^{-1/2} - \lambda = 0 \quad (6.43b)$$

$$\frac{\partial \Delta}{\partial \lambda} = -\frac{4}{25}(u_1 - 1)^2 - u_2 + 4 = 0 \quad (6.43c)$$

Eliminating λ and u_2 leaves a cubic equation which may be solved by trial and error to yield $(u_1, u_2) = (2.36, 2.19)$ from which $\beta = 3.22$ using Equation (6.41) and $p = 0.64 \times 10^{-3}$ as obtained from standard Normal tables.

6.8.5 Sub-systems and bounds

As already noted, for some elements or sub-systems there will be a number of 'failure' conditions, each leading to an 'unacceptable' outcome. The FOSM method and its derivatives cannot easily handle multiple limit state function problems. However it is possible to use so-called 'system bounds'.

Reliability of load-resistance sub-systems

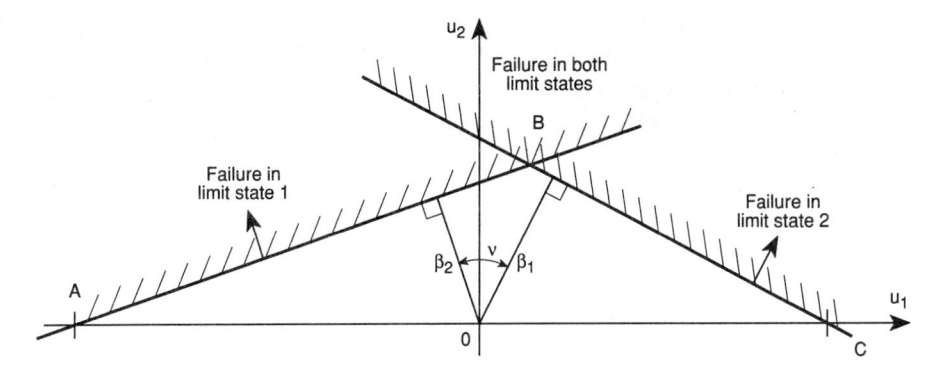

Figure 6.21 Intersection of limit state functions.

Consider, for example, the case with only two random variables and only two limit state functions, in standard Normal space, as shown in Figure 6.21. For a series system the 'unacceptable' region is defined by expression (6.38) above. The probability content in the region ABC is not easily calculated directly. Note, however, that the probability content in the region will be less than the product of the probability content for each of the two component regions. It will obviously be more than the sum of the probabilities of the two component regions, thus:

$$p_1 \cdot p_2 < p_{12} < p_1 + p_2 \qquad (6.44)$$

where each term is defined in Figure 6.21. It is possible to extend this argument intuitively to more than two dimensional space and to multiple limit state functions. In this case the probability content p_t for a region enclosed by a set of linear limit state functions is bounded by the combination of the (small) probabilities p_i for the n individual limit state functions (Cornell, 1967):

$$\max_{i=1}^{n} (p_i) \leq p_t \leq 1 - \prod_{i=1}^{n}(1 - p_i) \approx \sum_{i=1}^{n} p_i \qquad (6.45)$$

This result may be improved by considering the effect of intersecting regions (Kounias, 1968; Ditlevsen, 1979) to give:

$$p_1 + \max \left\{ \sum_{\substack{i=2 \\ j<i}}^{k \leq n}(p_i - p_{ij}) \right\} \leq p_t \leq \sum_{i=1}^{n} p_i - \sum_{\substack{i=2 \\ j<i}}^{n} \max(p_{ji}) \qquad (6.46)$$

In this expression the ordering of the limit state expressions (and hence the limit state associated with p_1) may influence the accuracy. Note that p_{ij} represents the probability content within a typical intersection region such as ABC in Figure 6.21. For calculations involving only second-moment ideas, this intersection probability can be obtained by simple numerical integration of the bi-variate Normal distribution (Owen, 1956; Johnston and Kotz, 1972). Alternatively, the intersection probability can be bounded (Ditlevsen, 1979).

The second-order bounds (6.46) are generally better than the first-order bounds (6.45) but considerable differences between the upper and the lower bounds may remain, particularly where the probabilities involved are not very small and where there is rather high correlation (dependence) between the limit state functions.

6.8.6 First Order Reliability method

For many problems the accuracy of the available information is such that the FOSM approach will be all that is warranted. In some problems, however, the probability distribution of one or more of the basic random variables may be rather different from Normal and the analyst may wish to take this into account. This can be achieved using the basic ideas of the FOSM method.

Recall that the region of interest for the evaluation of probabilities is the 'unacceptable' region, which is usually well away from the origin in standard Normal space. Hence the region of most interest is the 'tail' of the distribution. Consider now the non-Normal probability distribution shown in Figure 6.22. It could be approximated by an 'equivalent' Normal tail such that the probability content under the two tails is identical. Other conditions, such as the probability density at some point, say u^* (the 'design' or 'checking point'), being the same, could be imposed also. With this equivalence, the 'equivalent' Normal distribution would be used in the FOSM approach outlined in the previous section.

Since the checking point is not known at the outset, but is only obtained during the calculations, a trial value is required for obtaining equivalent Normal distributions. This means that the First Order Reliability method will be based on a trial and error process. When non-linear limit state functions are involved also, it is clearly not feasible to

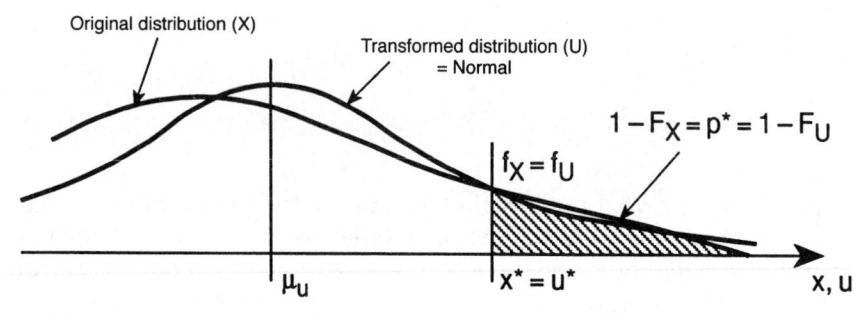

Figure 6.22 Original and transformed probability density functions.

carry out by hand the various calculations, except for rather trivial problems. As a result, various computer codes are now available for this task, as well as for second moment methods generally.

The first-order methods can be extended to allow for curvature of the limit state function in the calculation procedure, rather than ignoring the probability content lost by linearization. These are so-called 'second-order' methods. However, even more of the original simplicity associated with the first-order methods is then lost and numerical techniques are certainly required to deal with the computations. At this point consideration should be given to the use of Monte Carlo techniques.

6.8.7 Monte Carlo simulation methods

The First Order Second Moment method and the methods derived from it are often very useful. They do have the disadvantages, however, that

1. the 'design' or 'checking' point must be identified for each limit state function,
2. non-linear limit state functions are not easily handled and may give rise to inaccuracies,
3. for variables having non-Normal distributions both the random variables and the original problem must be transformed to the standard Normal space – a process which requires, in the general case, the use of the Rosenblatt (1952) Transformation.

It is the latter aspect which causes the most serious difficulty for practical problems involving non-Normal random variables and dependency.

As noted, a completely different approach to handling the integration of expression (6.37) is the use of Monte Carlo simulation methods. These can be formulated directly in the space of the original variables, so that no transformation is required, can handle any form of limit state function and are not restricted to Normal random variables. They suffer, in the minds of some analysts, from the disadvantage of high computer CPU times. However, with the increased availability of large capacity computers, this is increasingly less of an issue. Also, so-called 'variance reduction' techniques can be applied to reduce dramatically the number of simulation runs required for a given degree of output accuracy, as will be outlined.

(a) Simple (or crude) Monte Carlo methods

The multi-dimensional integral (6.37) can be re-written as (Rubinstein, 1981):

$$p_f = \int \ldots \int I[G(\mathbf{x}) \le 0] \cdot f_\mathbf{x}(\mathbf{x}) d\mathbf{x} \qquad (6.47)$$

where the indicator function $I[..]$ is defined such it takes on the value 1 if $[..]$ is true and zero otherwise. Equation (6.47) represents the first moment of $I[..]$. This means that an unbiased estimator of it is given by:

$$p_t \approx \frac{1}{N} \sum_{j=1}^{N} I[G(\hat{x}_j) \leq 0] \qquad (6.48)$$

where $\hat{x}_j$ represents the jth vector of realizations of the random variables, taken by random sampling from the joint probability density function $f_x()$. Equation (6.48) allows direct use of Monte Carlo techniques. A randomly generated vector such as $\hat{x}_j$ is placed in the limit state function expression $G_i()$ to evaluate whether the relevant limit state has been violated. If so, the sample represents a failure state and it is counted in the numerator of Equation (6.48). This process is repeated a sufficient number of times (N) to estimate the failure probability according to (6.48).

It should be clear that for highly reliable systems (i.e. p_t is very small), a very large number of simulation runs (i.e samples) will be required to obtain a reasonable number of them falling in the 'unacceptable' domain. Equivalently, the uncertainty (e.g. the variance) associated with the estimated value of p_t will be high unless the number of samples is large. It may be shown that the variance reduces with $N^{1/2}$. One particularly powerful technique for variance reduction will be described in the next section. Other techniques are available in the literature (e.g. Rubinstein, 1981; Melchers, 1993b).

(b) Importance sampling and related methods

Let Equation (6.48) be re-written as

$$p_t = \int \ldots \int I[G(x) \leq 0] \frac{f_x(x)}{h_v(x)} h_v(x) \, dx \qquad (6.49)$$

where now the function $h_v(\cdot)$ is known as the 'importance sampling probability density function'. It has the same dimensions as x. The probability of violating the limit state functions can now be estimated from:

$$p_t \approx \frac{1}{N} \sum_{j=1}^{N} \left\{ I[G(\hat{v}_j) \leq 0] \frac{f_x(\hat{v}_j)}{h_v(\hat{v}_j)} \right\} \qquad (6.50)$$

where $\hat{v}_j$ is a vector of sample values taken from $h_v(\cdot)$ (rather than from $f_x(\cdot)$). Note that $\hat{v}_j$ is defined in the space x – usually they will be identical spaces.

The reason for using $h_v(\cdot)$ is that if it is sensibly chosen, there can be a dramatic improvement in the accuracy with which p_t is estimated for a given number of samples. (Note that the converse also applies – a poorly chosen $h_v(\cdot)$ can lead to much worse results than obtained from

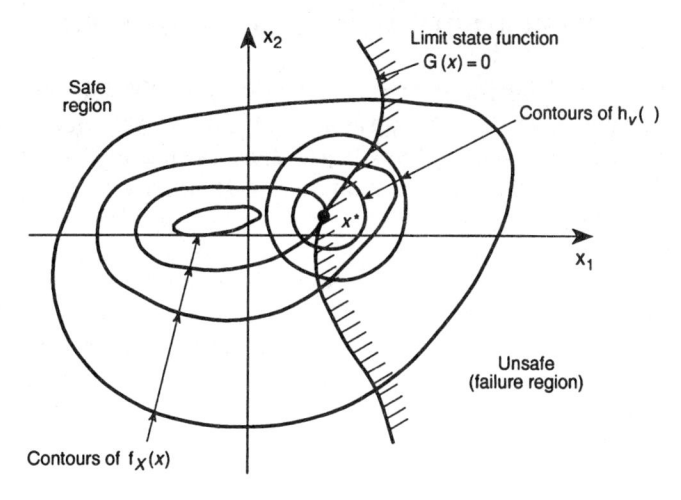

Figure 6.23 Importance sampling function h_v – in original space.

the conventional Monte Carlo method.) The trick is to select a $h_v(\cdot)$ such that the most information is extracted from the vector of samples $\hat{v}_j$. This means that $h_v(\cdot)$ should be selected to be in the area of interest for the integration in (6.37), that is, it should be such that it is located within the 'unacceptable' or failure region. This is illustrated schematically in Figure 6.23 for a simple limit state function and a two-dimensional random vector space.

Apart from the general rule that the ideal $h_v(\cdot)$ is topologically similar to the shape of $f_x(\cdot)$ in the region of integration, there are no rules. The selection is in the hands of the analyst. For some simple problems it has been suggested that an appropriate strategy is to select the 'design' or 'checking' point as the mean point for $h_v(\cdot)$ and 1–2 times the variance of $f_x(\cdot)$ as appropriate for the variance for $h_v(\cdot)$, assumed to be a multi-Normal distribution (Engelund and Rackwitz, 1993). However, this requires knowledge or an estimate of the checking point location. For some problems this location might, with sufficient accuracy, simply be estimated. For other problems recourse can be had to a trial and error approach, in which the function $h_v(\cdot)$ is updated and modified continually, according to the results of earlier samples – in the manner of a search algorithm. Computer routines for this purpose are available and further refinements are discussed by e.g. Melchers (1993b). In all cases the amount of computation required for a given accuracy in output is considerably below that required for simple Monte Carlo analysis.

6.9 SENSITIVITY ANALYSIS AND UPDATING

A sensitivity analysis is a way of ascertaining how the calculated output from a risk analysis might change as a result of a given change in the value of a particular input parameter. Clearly, sensitivity analysis gives a measure of the amount of effort which might be expended in refining the input value, whether through increased data collection and/or better interpretation, or through greater attention to event and/or system modelling. It should be used routinely and extensively in a systems risk analysis, but often is not. The most common sensitivity analysis approach is with point estimates, rather than with the probabilistic parameters of the inputs.

Sensitivity analysis can be undertaken also in conjunction with Second Moment and full distributional analyses. Evidently, if there is uncertainty about one (or more) input value(s), sensitivity analysis will assist in indicating the effect of the uncertainty; for example, in the effect of a change to the variance of one of the random variables.

The basic idea of a sensitivity analysis can be presented as follows. Consider the event tree shown in Figure 6.6, with the mean values only being of interest (i.e. the first number in each bracketed pair). Let there be uncertainty about the mean value for the probability of the fusible link functioning on demand. Consider a change in value from Yes = 0.99 to 0.9, that is a 10 % decrease. The consequent change in the SJF output is then found to be from 2.97 E-7 and 2.7 E-12 for SJF to 2.7 E-7 and 2.7 E-11 respectively. The total probability of occurrence of a SJF from this branch is then (3.0 E-7 + 2.7 E-11) = 3.0 E-7. This shows that the sensitivity of the occurrence of a SJF to a -10% change in the effectiveness of the fusible link is negligible. Similar analyses can be carried out for the other parameters in Figure 6.6, such as the effectiveness of the RSD, the effectiveness of the EFV etc.

There are various reasons why a risk analysis might need updating. For example, as a facility ages, the risk of component failure may change and this may have an influence on the outcome of the analysis. Also, with time, more experience might be gained about the failure rate of components. Such information may come from the facility itself, or from other sources. In addition, understanding of the functioning of the system could have been improved. For these and other reasons it may be desirable to update the risk analysis. In the nuclear industry the need for such updating has led to the concept of a 'living' risk analysis or probabilistic safety analysis (PSA).

In principle, updating of a risk analysis simply requires the reworking of the analysis with updated information. Most of this will be available at the component or sub-system level. For these there may be generic and historical data with which the original risk analysis was performed, and newly acquired observational data. Such different data sources can be combined using Bayesian analysis.

6.10 SUMMARY

System evaluations fall broadly into two categories; both are important for realistic, complex systems. The first deals with calculation techniques applicable to fault trees and to event trees and has each event described by an historical or subjective failure rate or a probability estimate. These are directly available for use in the analysis. The second deals with methods applicable to situations where event outcome probability estimates must be derived from load-resistance characteristics of the subsystem or component.

A risk analysis should account for all the dependencies between the events of which the system is composed. In general the interactions between the components is more complex than is the case for independent components. Dependency has caused difficulties in risk analysis procedures, particularly for fault trees and where the modelling of the system is too simplified.

Remarks are made also about the determination of the time to 'first exceedence' (i.e. failure) and the mean time to failure (MTTF), essentially equivalent terms but usually applied in different settings. Comments are also made about the effect of time on a risk analysis and it is noted that a risk analysis will need to allow for better understanding and refined data if it is to have ongoing validity and usefulness.

REFERENCES

Ansell, J. and Walls, L. (1995) A generic dependency model. In I.A. Watson and M.P. Cottam (eds), *European Safety and Reliability Conference, ESREL '95*, The Institute of Quality Assurance, UK, 695–705.

Apostolakis, G. and Yum Tong Lee (1977) Methods for the estimation of confidence bounds for the top-event unavailability of fault trees, *Nuclear Engineering and Design*, **41**, 411–19.

Apostolakis, G. (1987) Uncertainty on probabilistic safety assessment. In F.H. Wittmann (ed.), *Structural Mechanics in Reactor Technology*, Vol. M, A.A. Balkema, Rotterdam, 395–401.

Berger, J.O. (1985) *Statistical Decision Theory and Bayesian Analysis*, 2nd edn, Springer-Verlag, New York.

Cornell, A.C. (1967) Bounds on the reliability of structural systems, *J. Struct. Engineering*, ASCE, **93**(ST1), 171–200.

Cornell, A.C. (1982), Some thoughts on systems and structural reliability, *Nuclear Engineering and Design*, **71**, 345–8.

Der Kiureghian, A. and Moghtaderi-Zadeh, M. (1982) An integrated approach to the reliability of engineering systems, *Nuclear Engineering and Design*, **71**, 349–54.

Ditlevsen, O. (1979) Narrow reliability bounds for structural systems, *J. Struct. Mech.*, **7**(4) 453–72.

Doerre, P. (1989) Basic aspects of stochastic reliability analysis for redundant systems, *Reliab. Engineering System Safety*, **24**, 351–75, 381–5.

Dougherty, E.M. and Fragola, J.R. (1988) *Human Reliability Analysis*, Wiley, New York.
Engelund, S. and Rackwitz, R. (1993) A benchmark study on importance sampling techniques in structural reliability, *Structural Safety*, **12**(4), 255–76; **14**, 299–302.
Farmer, F.R. (1982) Decision in reliability analysis, *Nuclear Engineering and Design*, **71**, 399–403.
Fleming, K.N., Mosleh, A. and Deremer, R.K. (1986) A systematic procedure for the incorporation of common cause events into risk and reliability models, *Nuclear Engineering and Design*, **93**, 245–73.
Green, A.E. and A.J. Bourne (1972) *Reliability Technology*, London, John Wiley & Sons.
Greig, G.L. (1993) Second moment reliability analysis of redundant systems with dependent failures, *Reliability Engineering and System Safety*, **41**, 57–70.
Hollnagel, E. (1993) *Human Reliability Analysis: Context and Control*, Academic, London.
International Study Group (1985) *Risk Analysis in the Process Industries*, Rugby, The Institution of Chemical Engineers.
Jackson, P.S. (1982) A second-order moments method for uncertainty analysis, *IEEE Trans. on Reliability*, **R-31**(4), 382–8.
Johnston, N.L. and Kotz, S. (1972) *Distributions in Statistics: Continuous Multivariate Distributions*, John Wiley, New York.
Kafka, P. and Polke, H. (1986) Treatment of uncertainties in reliability models, *Nuclear Engineering and Design*, **93**, 203–14.
Kaplan, S. and Garrick, B.J. (1981) On the quantative definition of risk, *Risk Analysis*, **1**, 11–37.
Kelly, B.E. and Lees, F.P. (1986) The propagation of faults in process plants: 1, Modelling of fault propagation, *Reliability Engineering*, **16**, 1–38.
Kirwin, B. (1994) *A Guide to Practical Human Reliability Assessment*, Taylor & Francis, London.
Kounias, E.G. (1968) Bounds on the probability of a union, with applications, *Amer. Math. Stat.*, **39**(6), 2154–8.
Laviron, A. (1985) Error transmission in large complex fault trees using the ESCAF method, *Reliability Engineering*, **12**, 181–92.
Lipow, M. (1982) Number of faults per line of code, *IEEE Transactions on Software Engineering*, **SE-8**(4), 437–9.
Madsen, O.H., Krenk, S. and Lind, N.C. (1986) *Methods of Structural Safety*, Prentice-Hall Inc., Englewood Cliffs, New Jersey.
Mazumbar, M. (1982) An approximate method for computation of probability intervals for the top-event probability of fault trees, *Nuclear Engineering and Design*, **71**, 45–50.
Melchers, R.E. (1987) *Structural Reliability Analysis and Prediction*, Chichester, Ellis Horwood/John Wiley & Sons.
Melchers, R.E. (1993a) On the treatment of uncertainty information in PRA. In R.E. Melchers and M.G. Stewart (eds), *Probabilistic Risk and Hazard Assessment*, Balkema, Rotterdam, 13–26.
Melchers, R.E. (1993b) Modern computational techniques for reliability estimation. In K.S. Li and S-C.R. Lo (eds), *Probabilistic Methods in Geotechnical Engineering*, Balkema, Rotterdam, 153–63.
Melchers, R.E. and Tang, L.K. (1984) Dominant failure modes in stochastic structural systems, *Structural Safety*, **2**, 127–43.
Mosleh, A. (1991) Common cause failures: an analysis methodology and examples, *Rel. Engineering and System Safety*, **34**(3), 249–92.

Owen, D.B. (1956) Tables for computing bivariate, Normal probabilities, *Ann. Math. Stats.*, **27**, 1075–90.
Parry, G.W. (1989) Discussion of Doerre, P., 'Basic aspects of stochastic reliability analysis for redundant systems', *Reliab. Engineering System Safety*, **24**, 377-81.
Porn, K. and Shen, K. (1992) On the integrated uncertainty analysis in probabilistic safety assessment. In K.E. Petersen and B. Rasmussen (eds), *Proc. European Safety and Reliability Conf. '92 (ESRC '92)*, Elsevier Applied Science, London, 463–75.
Qin Zhang, (1989) A general method dealing with correlations in uncertainty propagation in fault trees, *Reliability Engineering and System Safety*, **26**, 231–47.
Ragheb, M. and Abdelhai, M. (1987) Uncertainty propagation in fault trees using quantile arithmetic methodology. In F.H. Wittmann (ed.), *Structural Mechanics in Reactor Technology*, Vol. M, A.A. Balkema, Rotterdam, 403–9.
Rigby, L.V. and Swain, A.D. (1968) Effects of assembly error on product acceptability and reliability, *Proceedings of the 7th Annual Reliability and Maintainability Conference*, ASME, New York, 312–19.
Rosenblatt, M. (1952) Remarks on a multivariate transformation, *Ann. Math. Stat.*, **23**, 470–2.
Rubinstein, R.Y. (1981) *Simulation and the Monte Carlo Method*, John Wiley, New York.
Shooman, M.L. (1968) *Probabilistic Reliability – An Engineering Approach*, McGraw-Hill Book Co.
Smith, D.J. (1985) *Reliability and Maintainability in Perspective*, 2nd edn, London, Macmillan.
Stewart, M.G. (1993) Structural reliability and error control in reinforced concrete design and construction, *Structural Safety*, **12**, 277–92.
Swain, A.D. (1963) *A Method for Performing a Human Factors Reliability Analysis*, Monograph SCR-685, Sandia National Laboratories, Albuquerque, New Mexico.
Swain, A.D. and Guttman, H.E. (1983) *Handbook of Human Reliability Analysis with Emphasis on Nuclear Power Plant Applications*, NUREG/CR-1278, US Nuclear Regulatory Commission, Washington, DC.
Thoft-Christensen, P. and Murotsu, Y. (1986) *Application of Structural Systems Reliability Theory*, Springer Verlag, Berlin.
Turkstra, C.J. (1970) *Theory of Structural Design Decisions*, Study No. 2, Solid Mechanics Division, University of Waterloo, Waterloo, Ontario.
Watson, I.A. and Johnston, B.D. (1987) Treatment of CMF / CCF in PSA. In F.H. Wittmann (ed.), *Structural Mechanics in Reactor Technology*, Vol.M, A.A. Balkema, Rotterdam, 125–40.

CHAPTER 7

Risk acceptance criteria

7.1 INTRODUCTION

Chapters 3–6 have dealt with risk analysis procedures and the estimation of system risk. The next phase of risk management (see Chapter 1) is risk assessment. It might be defined as a 'decision-making process whereby a level of risk is compared against criteria and risks are prioritized for action' (AS/NZS 4360, 1995). For societal risk, the responsibility for these decisions is normally conferred on one or more decision-makers; typically decision-makers tend to be regulatory agencies or authorities. In private enterprise the decision-maker is usually company management. They must decide, for instance, whether to proceed with the development, modification or expansion of new or existing systems. Regulatory authorities are mostly concerned with plant personnel and public health, safety and environmental issues. These are the matters normally considered when deciding whether to issue development consent, system certification, a permit or licence to operate. Company management may be more concerned about discretionary requirements such as setting priorities to ensure the viability, efficiency, cost-effectiveness and safety of their system.

In its broadest perspective, risk assessment is concerned with a number of issues: these are often intertwined with political processes. Matters to be addressed include:

> Who is to bear what level of risk, who is to benefit from risk-taking and who is to pay? Where is the line to be drawn between risks that are to be managed by the State, and those that are to be managed by individuals, groups or corporations? Where is the line to be drawn between minimizing accidents by anticipation and promoting resilience to cope with whatever failures may arise, and between attempts to influence the causes of hazards as against measures to change their effects? What information is required for 'rational' risk management and how should it be analysed? What actions make what difference to risk outcomes? Who evaluates success or failure in risk management and how? Who decides what should be the desired trade-off between different risks? (Hood *et al.*, 1992).

These are very complex issues and are seldom tackled using risk assessment alone. The decision-making process will be influenced by

Introduction

political processes, but for normal situations is likely to be influenced by, among other things:

1. anticipation of system failure and resilience against unexpected catastrophe;
2. assumptions used to compute a numerical estimate of system risk;
3. size of uncertainties in estimating system risks (e.g. some regulatory safety targets may be inappropriate for system risks with large uncertainties);
4. organizational vulnerabilities to system failure (e.g. safety culture);
5. cost of risk reduction;
6. size and composition of groups involved in decision-making process;
7. aggregation of individual preferences (i.e. distribution of benefits and risks); and
8. counter-risks (alternatives may have other societal risks).

In principle, these issues are all related to risk acceptance criteria and hinge on the question; 'what risks are acceptable?' This question forms the focus for the present chapter, which will consider the processes associated with the development of risk acceptance criteria. This involves:

1. perception of risk: ensuring that the level of system risk as perceived is acceptable (or tolerable), especially for societal risks;
2. formal decision analysis: analytical techniques to balance or compare risks against benefits (e.g. risk-cost-benefit analysis); and/or
3. regulatory safety goals: legislative and statutory framework for the development and enforcement of risk acceptance criteria

In the following discussion particular emphasis is placed on the use of risk acceptance criteria in the area of regulatory requirements. The methods described may also be used by management to develop risk-based discretionary requirements – however this issue is not addressed specifically.

The entity charged with making the decision may be an individual in isolation, an individual representing one or a number of institutions or it may be a collective group, such as a committee. In all cases, 'good' collective decisions are generally considered to be those which have the following features (Paté-Cornell, 1984):

1. the preferences of all interested parties are expressed completely, freely and as accurately as possible;
2. those who are likely to be most affected by a decision should have the greatest influence on the decision (i.e. decisions are equitable where risks and benefits are shared); and

3. ideally a decision should be pareto optimal (i.e. no individual should suffer a net adverse effect).

In addition, good collective decisions will also be those which have stood the scrutiny of quality assurance measures and independent review of the decision-making processes. Nevertheless, it will be acknowledged that who the decision-maker is may have an influence on the decision-making process and on the decision outcome: equally, it should be recognized that such decisions may be overruled (or at least delayed) by such political considerations as electoral pressure, national security implications, or lack of funds.

7.2 DECISION-MAKERS AND SOCIETY

Decisions related to large-scale risks are made by a variety of private and public institutions (or organizations) because it is usually the case that 'individuals in crises do not make life and death decisions on their own. Who shall be saved and who shall die is settled by institutions' (Douglas, 1987). Private and public institutions involved in risk management typically include national, state or local governments; public regulatory authorities; insurance companies; private companies; and lobby groups. Table 7.1 shows the types of institutions which may be involved in public risk management. The institutions may well have conflicting preferences; usually some of these may be resolved through bargaining. Other areas of conflict may not be so easily resolved owing to political, territorial and functional differences. The resources available to an institution (and how these are used) will also affect the role the institution may play in the decision-making process. Resources include the availability of data (for predicting and forecasting), financial

Table 7.1 Examples of institutions involved in public risk management

	Institutional type		
Territorial Level	Core executive bodies	Independent public bodies	Private or independent bodies
International	EU Commission	EU Court of Justice	Greenpeace
National	National parliaments	National courts and independent regulatory bodies	National Association of Insurers
Sub-national	State or local governments	Independent regional/local statutory bodies	Local firms and activists

Source: adapted from Hood et al. (1992).

resources (i.e. adequacy of funding), regulatory capacity, and measures available (and used) to ensure compliance with regulations (Hood et al., 1992).

Most systems are subject to review by regulatory authorities. These include national bodies such as the UK Health and Safety Executive, and the UK Nuclear Installation Inspectorate, federal bodies such as the US Nuclear Regulatory Commission, the US Environmental Protection Agency, and the US Food and Drug Administration, and state-based bodies such as the Department of Planning (NSW). Regulatory authorities are created as a result of government legislation. Usually they have a statutory obligation to take into account the preferences of all interested parties. However, this may not be feasible; for instance, there are approximately 4000 public organizations and lobby groups in Canada alone (Delbridge, 1990). To overcome this problem, it is usually the case that, prior to adopting a regulation, the regulatory authority must

1. publish the proposed regulation,
2. provide reasons why it should be adopted, and
3. provide an administrative procedure for the collection and review of public submissions.

The response to the public submissions is then published together with the final regulations. In this way, public participation in the decision-making process may help offset institutional biases. It may also provide a greater degree of acknowledged legitimacy to the regulation. A roughly parallel process of public participation is in place in many countries for approval of construction of new or modified engineering systems. Despite these procedures, it is clear that regulatory actions cannot address equally well all public concerns, because to do so may cause conflict with societal goals or because some actions have major economic and social dislocations (as might be incurred, for example, through closure of the facility, and hence through loss of employment).

Usually, an independent appeal mechanism exists. It is a review, sometimes judicial, which determines whether the regulatory authority has complied with its statutory or legislative requirements. The review is an attempt to ensure that opposing views are not be omitted from the decision-making process. In difficult (political, social) cases, a public or parliamentary inquiry (e.g. Royal Commission headed by a senior judge) may be deemed necessary to resolve controversial or political considerations and also to restore public confidence in the decision-making process. Tann (1992) provides an interesting description of a 1986 public inquiry lasting 95 days held to assess a planning application for a nuclear reprocessing plant in Scotland.

7.3 RISK PERCEPTION

7.3.1 Acceptable and tolerable risk

Normally, it is highly desirable that the estimated levels of risk associated with a facility or to which the facility contributes are below those levels of risk deemed to be just acceptable (or at least tolerable) to society. This raises the very difficult question of what level of risk is tolerable to society. It depends, generally, on the values, beliefs and attitudes of society. It is expected that societal risk perception will be influenced by psychological, sociodemographic and other variables; hence the perception and acceptability of risk will vary from community to community. For this reason, it is unlikely that a single numerical estimate of system risk (i.e. regulatory quantitative safety target) can represent the accepted risk for society as a whole. Nonetheless, ensuring that system risks are below regulatory safety targets is often an important, but not sole, criterion in the decision-making process. The importance of such a criterion may diminish if system risks are significantly lower than the regulatory safety targets. This will be discussed later in section 7.5.

The term 'acceptable risk' may not represent realistically the views of an individual or of society because some risks may never be perceived as being acceptable. On the other hand, a risk might be tolerable, reflecting the extent to which an individual or society is just prepared to tolerate (albeit with some reluctance) that risk. For example, a risk may be tolerable if the benefits appear to exceed the risk. The Health and Safety Executive in the UK defines tolerable risk as:

> Tolerability does not mean acceptance. It refers to the willingness to live with a risk to secure certain benefits and in the confidence that it (risk) is being properly controlled. To tolerate a risk means that we do not regard it as negligible or something we might ignore, but rather as something we need to keep under review and reduce still further if and as we can (HSE, 1988).

It is clear that technological and social aspects both need be considered in the setting of acceptable risk criteria. To avoid confusion over terminology it is assumed that the definition of 'acceptable risk' also includes within it the above definition of tolerable risk; namely, an acceptable risk is also tolerable.

At precisely what level a risk may be deemed acceptable (or tolerable) is uncertain; however, how the risk is perceived is an important factor that will influence risk acceptability. Pidgeon *et al.* (1992) suggests that perceived risk is influenced by a combination of:

1. objective risk;
2. cognitive psychology;

3. social, cultural and institutional (e.g. political) processes; and
4. risk communication.

These aspects of risk perception are now described.

7.3.2 Objective risk

An objective risk is an estimate of system risk obtained from known statistics (e.g. annual expected fatalities from car accidents) or from quantified risk analyses methods, such as QRA and PRA (see Chapter 6). Objective risks may be expressed in a variety of forms such as fatalities per year, fatalities per year of exposure, lifetime fatalities, fatalities per hour of exposure divided by the number of people exposed multiplied by 10^8 (the so-called Fatality Accident Rate, FAR), and so on. A distinction must be made between individual and societal risks. A societal risk is the probability that the consequences will affect more than one individual. It is typically represented in the form of an F-N curve, which is a plot of cumulative frequency versus consequences (see Figure 7.1). The magnitude of the objective risk and the terms in which it is expressed will obviously affect risk perception. For example, an individual fatality risk of 10^{-4} is equivalent (in a statistical sense) to a societal risk of 10^{-6} of killing 100 people. Rather illogically, but perhaps understandably, society generally appears to be more concerned about catastrophic events that harm many people than a series of lesser failure events that collectively harm a similar number of people (so-called 'risk aversion', see section 7.4.3).

In this regard, it should be recognized that estimates of system risk obtained from formal risk analyses may be not as objective as sometimes thought. Risk analyses are dependent upon the attitudes, judgement and experience of the analysts; they affect the scope of the study, the choice of risk/consequence models, the treatment of uncertainties, incorporation of human reliabilities, etc. Thus some computed 'objective' risks may be less realistic than risks estimated from statistical inference on known failure events.

7.3.3 Psychological aspects

A psychological view of risk perception is concerned with the way external information about hazards is selected for attention and interpreted by an individual. A useful way to assess an individual's perception of risk is to ask individuals to make fatality estimates (i.e. expected number of fatalities per year) for a range of hazards. These estimates can then be compared with objective fatality statistics. In one such study, Lichtenstein *et al.* (1978) found that individuals overestimated deaths resulting from infrequent causes (e.g. tornadoes, accidents, floods) and underestimated deaths resulting from frequent causes (e.g. cancer,

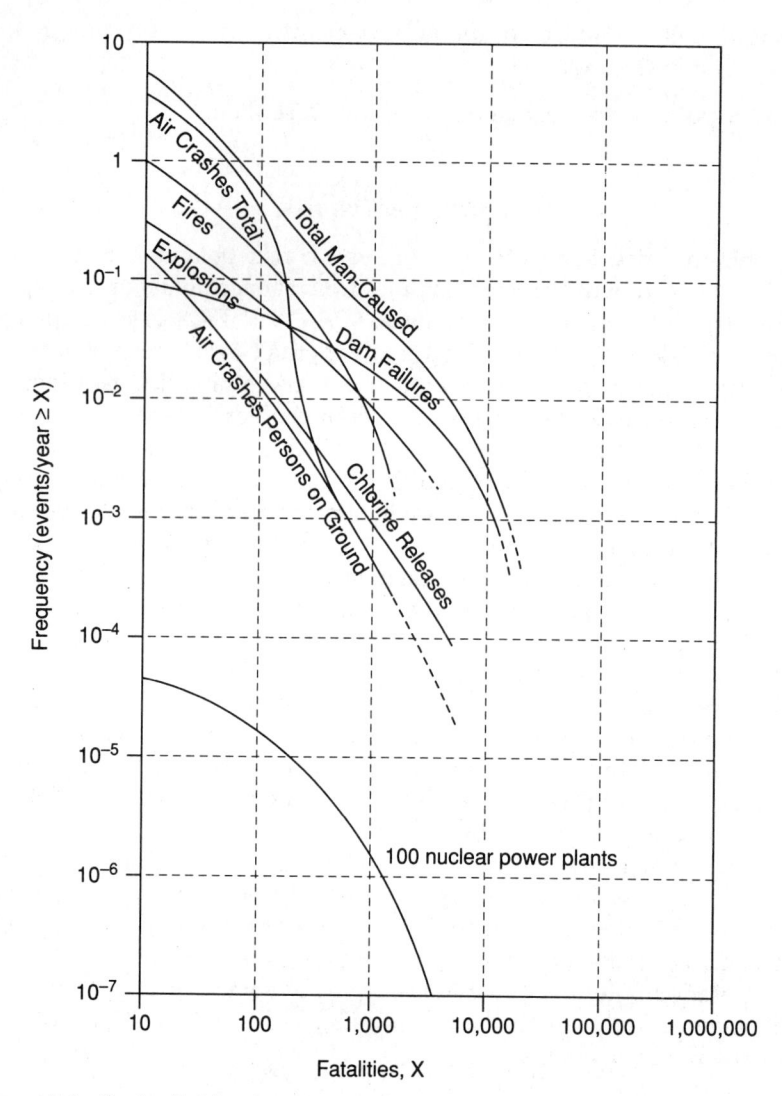

Figure 7.1 Societal risk of man-caused events involving fatalities.
Source: USNRC (1975).

diabetes, stroke). It was found also that individuals overestimated expected deaths for easily recalled, vivid or imaginable causes of death. While the absolute estimates of perceived and objective expected deaths did vary, it was observed by Slovic *et al.* (1980) that the ranking of comparative risks was not affected significantly. On the other hand, Brehmer (1994) and others suggest that 'laymen's judgements of risk are made in a way that is almost totally unrelated to the kinds of concepts that enter into the estimates made by engineers and statisticians'.

Table 7.2 Comparison of objective and perceived risks

Rank[a]	Activity/technology	Perceived risk[b]	Fatality estimates[c]	Risk[d] (×10⁻⁶ fatalities/person/year of exposure)
1	Nuclear weapons	78	–	–
2	Warfare	78	–	–
4	Handguns	76	17 000	
5	Crime	73		93
6	Nuclear power	72	100	9
9	Smoking	68	150 000	4000–22 000
10	Terrorism	66		0.9
13	Nerve gas	60	–	–
15	Alcoholic beverages	57	100 000	
17	Motor vehicles	55	50 000	3000–6100
31	Fire fighting	44	195	
32	Motorcycles	43	3000	
36	DNA research	41	–	–
49	Dams	31	30	
51	Commercial aviation	31	130	4000–11 000
61	Railroads	29	1950	400–700
64	Mountain climbing	28	30	300 000
65	Bridges	27		
69	Skyscrapers	26		0.2
70	Electric power	26	14 000	
73	Downhill skiing	26	18	9000
74	Space exploration	25	–	–
90	Solar electric power	12	–	–

Note: [a] ranking of perceived risks (most risky to least risky);
[b] risks rated on a 0–100 scale (from 'not risky' to 'extremely risky');
[c] annual expected fatalities obtained from existing statistics (US);
[d] adapted from Melchers (1987); Cox, et al. (1992); and Atallah (1980).

Individual attitudes and responses to risk are generally more complex than responses obtained for expected deaths (e.g. Pidgeon et al., 1992). For example, Table 7.2 shows the rankings of perceived risk for some activities and technologies and the 'objective' expected number of fatalities per year. It is seen that nuclear power plants have one of the highest perceived risks and the lowest expected number of annual fatalities. Slovic et al. (1980) concluded that this perception about nuclear power plants is primarily influenced by their 'catastrophic potential' because 65% of respondents expected more than 10 000 fatalities if next year was a 'disastrous year'. The study concluded that an individuals' negative attitude to risk is influenced (in part) by the following qualitative risk factors:

1. 'dread' risk – uncontrollability, dread (or fear), involuntary, catastrophic potential, little preventative control, certain to be fatal, inequitable

distribution of risk, threatens future generations, affects me personally, risk not easily reduced, and risks increasing (e.g. chemical weapons, terrorism);
2. 'unknown' risk – not observable, unknown to those exposed, effects immediate, new (unfamiliar), and unknown to science (e.g. genetics research, space exploration, food irradiation); and
3. high degree of exposure – large number of people exposed to hazard (e.g. alcoholic beverages, caffeine, herbicides).

Otway and von Winterfeldt (1982) have suggested similar qualitative risk factors, see Table 7.3. Green and Brown (1978) proposed that the severity of an injury or disability ('fate worse that death', 'the living dead') is also a significant qualitative risk factor. Risk factors for a specific hazard were found to be highly correlated within each category; for example, risks with catastrophic potential were also judged as dread risks. Correlations were low between risk factors from different categories; the exception to this was nuclear power plants which rated high in all three risk categories. The above qualitative risk factors may explain in part why some individuals exhibit what appears to be contradictory behaviour. For example, smokers may be more concerned about an accident at a nuclear power plant than the effect of smoking on their own health (e.g. Lave, 1987). This difference in perceived risks most likely is influenced by the 'dread' and 'unknown' aspects of risk associated with nuclear power plants.

Rather than expressed views or behaviour patterns, another indicator of perceived risk is the amount of money spent to ameliorate the perceived risk. At the societal level, the expenditure per life estimated to be saved under specific government risk reduction programmes is

Table 7.3 General (negative) attributes of hazards that influence risk perception and acceptance

Involuntary exposure to risk
Lack of personal control over outcomes
Uncertainty about probabilities or consequences of exposure
Lack of personal experience with the risk (fear of unknown)
Difficulty in imagining risk exposure
Effects of exposure delayed in time
Genetic effects of exposure (threatens future exposure)
Infrequent but catastrophic accidents
Benefits not highly visible
Benefits go to others
Accidents caused by human failure rather than natural causes

Source: adapted from Otway and von Winterfeldt (1982).

Table 7.4 Money spent to save a life, for government risk reducing regulations

Regulation	Annual cost ($ million)	Annual lives saved	$ million per life saved
Steering column protection	130	1300	0.1
Unvented space heaters	6.3	63	0.1
Cabin fire protection	3	15	0.2
Passive restraints/belts	555.0	1850	0.3
Alcohol and drug control	2.1	4.2	0.5
Seat cushion flammability	22.2	37	0.6
Floor emergency lighting	3.5	5	0.7
Hazard communication	360	200	1.8
Radionuclides/uranium mines	7.6	1.1	6.9
Arsenic/glass plant	2.2	0.11	19.2
Arsenic/copper smelter	1.6	0.06	26.5
Uranium mill tailings/active	111.3	2.1	53.0
Asbestos	6670.7	74.7	89.3
Arsenic	1082.3	11.7	92.5
DES (cattlefeed)	8976	68	132.0

Source: adapted from Lind *et al.* (1991).

based on the hypothesis that the greater the perceived risk the greater will be the money spent to reduce that risk. For a range of hazards, Lind *et al.* (1991) estimated the ratio of the annual costs (borne by consumers, industry etc.) resulting from US government regulations for risk reduction and the number of expected lives saved for a range of hazards, see Table 7.4. It appears that society (as represented by the US government) is prepared to spend more money to prevent death caused by exposure to toxic substances (asbestos, arsenic, radiation) than it is to prevent fatalities caused by motor vehicle accidents. This observation tends to confirm the importance of the qualitative risk factors described above.

7.3.4 Social, cultural and institutional processes

Risk perception for individuals and groups is affected by socio-demographic variables such as culture, nationality, gender, age and employment. The long-term psychological predisposition of the individual (e.g. risk taking, risk aversion) is also an important consideration. It is therefore expected that differences in risk perception will occur between individuals and between groups. Some individuals and groups may approve of the use of 'rational' or formal decision analyses if they believe in the concept (or world view) of a high-growth, high-technology society. However, others may be more concerned about environmental and social issues; hence, they would prefer the use of more participative decision-making methods. It is important to recognize that these differing views may offer new insights of the problem (and its solution).

The influence of culture is illustrated by the widespread acceptance of nuclear power in France and its rejection in the US. France has a traditional respect for technocrats and a strong central government, so there is limited public participation in decision-making. On the other hand, US citizens perhaps have rather less trust in technology and governments and therefore want more public participation and control over decisions (e.g. Morone and Woodhouse, 1989; Slovic, 1995).

7.3.5 Risk communication

It has been argued in section 7.3.3 that the perception of risk is influenced by the knowledge that the decision-maker has about hazards and risk. This knowledge may be 'ill-informed', 'well-informed' or somewhere in-between. Risk communication is the means through which information about hazards and risks are obtained, it is not simply about the education of the public (Rohrmann, 1995). The risk communication process consists of one or more of the following components:

1. instructions issued from 'experts' to the public (i.e. one-way transmission of information);
2. free exchange of information between interested parties; and
3. interpretation of actions (i.e. non-verbal communication).

The information about risks communicated by these components may or may not resolve differences of opinion, much less conflicts, between interested parties. Nonetheless, risk communication is essential if all are to participate fully in the decision-making process. It is often a legislative requirement that public and private institutions (1) inform the population about the hazards to which they may be exposed and (2) prepare emergency procedures in the event of system failure (Pidgeon et al., 1992). However, the known existence of emergency procedures may cause a 'reassurance-arousal' paradox, where the knowledge of preparations for an emergency may reassure the exposed population, while at the same time it may arouse fears in the same population because the facility may be perceived to be more hazardous than had been assumed previously (Otway and Wynne, 1989) (a phenomenon not unknown to commercial airline passengers).

Lack of trust between the interested parties inhibits the reconciliation of viewpoints. For instance, it is clear that trust in physicians makes surgical procedures acceptable, whereas lack of trust makes the management of potentially hazardous industries suspect (e.g. Slovic, 1995). Further, information may well be misused (e.g. results quoted out of context) by some participants in order to influence the preferences of other participants in the decision-making process.

Finally, Lichtenstein et al. (1978) found that newspaper articles concentrated almost entirely on catastrophic events (tornadoes, homicides,

Table 7.5 Some formal definitions of risk

Probability of undesired consequences.
Seriousness of (maximum) possible undesired consequences.
Multi-attribute weighted sum of components of possible undesired consequences.
Probability × seriousness of undesired consequences (expected loss).
Probability weighted sum of all possible undesired consequences (average expected loss).
Semivariance of possible undesired consequences about their average.
Variance of all possible consequences about mean expected consequences.
Weight of possible undesired consequences (loss) relative to comparable possible desired consequences (gain).

Source: adapted from Pidgeon *et al.* (1992).

motor vehicle accidents) while many diseases were rarely reported, even though annual fatalities from all diseases is approximately two orders of magnitude greater than that for homicides. It is not unreasonable to conclude that 'much of the information to which people are exposed provides a distorted picture of the world of hazards' (Slovic *et al.*, 1980).

7.3.6 Discussion

The preceding discussion has concentrated on risks expressed in terms of expected fatalities and risk of death per person per year of exposure. However, there is a large number of definitions for 'risk' or 'riskiness' (see Table 7.5), particularly as related to consequences. The latter may refer to fatalities as well as injury, property damage and damage to the environment. It follows that the perception of risk will be related to some degree to the definition of risk used or assumed, and also that the definition of risk is likely to vary from individual to individual. This inconsistency must be allowed for in comparing different sources of information about risk and risk perception.

A number of times already comment has been made about the difference between 'perceived' and 'objective' risks. Both contain subjective input, since even 'experts' need to make subjective assessments in their analysis of objective risks. Fischhoff (1989) has suggested that any conflict between perceived risk and objective risk may be seen as a 'conflict between two sets of risk perceptions' – sometimes referred to as 'public' and 'expert' estimates of risk respectively. Neither objective nor perceived risks may be deemed the 'correct' criteria for determining an acceptable level of risk. What might be considered 'correct' depends on the situation. For example, the public is often the main bearer of risk; it is important, therefore, that the decision-making process considers public as well as expert risk preferences. It follows that lay perceptions of risk should not be dismissed simply as uninformed.

Given appropriate stimuli, the 'lay person' can become an 'expert' in a very short span of time, and their expertise can be all the more formidable because it combines formal technical knowledge and local knowledge that is as relevant as it is unstructured and informal (Jasanoff, 1993).

7.4 FORMAL DECISION ANALYSIS

7.4.1 Objectives and attributes

Formal decision analysis techniques provide decision-makers with analytical techniques to assess risk preferences: in particular to compare or balance risks against benefits. An optimal solution (chosen from among several alternatives) may be achieved by maximizing either (1) the expected monetary value (e.g. risk-cost-benefit analysis) or (2) the expected utility. These are essentially the same processes but formulated in monetary terms and utilities respectively. Each of these will be discussed briefly below; first, however, some comments applicable to both are appropriate.

As will be seen, maximizing the expected value or the expected utility requires the estimation of system risk; both probability of occurrence of consequences and the magnitude of the consequences themselves. In the case of expected value analysis the consequences (and other outputs and inputs) need to be expressed in an (agreed) value system (such as monetary units). For utility theory, the value system is rather different – it is in terms of the perceived utilities that persons or groups attach the particular outputs and inputs. These are more difficult to estimate than monetary values, but are considered to be capable of representing non-monetary items.

Formal decision analyses assume that decisions are attributed to a single (or unitary) decision-maker (e.g. Keeney and Raiffa, 1976). This is reasonable in many cases, but in others an organization might be involved, for example, a regulatory authority. In this case the 'individual' is referred to herein as a 'supra' decision-maker. Evidently, the preferences of a supra decision-maker will be influenced by the preferences of other interested parties.

A formal decision analysis requires the establishment of objectives and attributes; this involves:

1. identifying the interested parties;
2. defining the objective(s) of each interested party (e.g. minimize adverse consequences), with the objectives measured in terms of economic, environmental and social values; and
3. representing each objective as a single attribute, where the attribute

Table 7.6 Objectives and attributes for siting nuclear power plants

Attribute	Objective	Category	Interested parties[a]
X_1	Minimize pollution	Environmental	E (or L)
X_2	Provide aesthetically pleasing facilities	Environmental	L
X_3	Minimize human health hazards	Human safety	L (or E,P,S,F)
X_4	Provide necessary power	Consumer well-being	S
X_5	Minimize consumer power costs	Consumer well-being	S
X_6	Maximize economic benefits to local community	Economic	L
X_7	Maximize utility company profits	Economic	P
X_8	Maximize state revenues	Economic	S
X_9	Improve balance of payments	Economic	F
X_{10}	Reduce dependency on foreign fuels	National interest	F

[a] E = environmentalists; L = local communities; P = power company; S = state agency; F = federal agency.
Source: adapted from Keeney and Raiffa (1976).

provides a scale (either qualitative or ordinal) for measuring the degree to which its objective is met.

A large number of objectives and attributes may be required to describe adequately the preferences of all interested parties. For example, Table 7.6 shows some of the objectives and attributes that may be associated with the siting of a nuclear power plant. In this case it is very unlikely to be possible to maximize simultaneously the benefits (e.g. power company profits) and minimize the adverse effects (e.g. consumer power costs) of a decision involving all interested parties. The decision-maker is required to make concessions or tradeoffs between beneficial and adverse consequences.

Keeney and Raiffa (1976) have observed that many decision-makers are not undecided prior to the commencement of a formal decision analysis. In other words, it appears that many decision-makers have already decided what they believe the 'correct' decision should be. In such cases, the formal decision analysis is then used to provide psychological comfort (i.e. confirm the decision-makers' intuition), help the communication process and justify his or her conclusion to others. In this case it is clear that the formal decision analysis process cannot be entirely objective.

7.4.2 Expected value analysis

An expected value analysis is suitable only if all interested parties share the same objectives and attributes, such as when their objectives and atrributes can be expressed in a common value system such as money. The expected value (EV) is then calculated as

$$EV_k = \sum_{i=1}^{M} p_i \left(\sum_{j=1}^{N} X_{ji} \right) \quad (7.1)$$

where k is the alternative or system configuration being considered, i is the 'state of nature' (e.g. normal operation, loss of coolant, fire, earthquake, and other modes of system failure), p_i is the probability of occurrence for each 'state of nature' i, M is the number of states of nature, j is the attribute, X_{ji} is the attribute value associated with the occurrence of each state of nature (e.g. damage cost for loss of coolant accident) and N is the number of attributes. The optimum decision is then the one which will determine the outcomes (e.g. site location, additional safety systems, better training etc.) that satisfy the objectives; for example, maximize the expected value if all objectives are to maximize economic benefits. Note, also, that in the above formulation the p_i values are obtained from a risk analysis, as are some of the values of X_{ji} (i.e. where these are consequences).

The most widely used procedure for expected value analysis is risk-cost-benefit analysis. In this case all attributes are given in monetary terms and money is used to represent all considerations associated with a decision. Thus the attributes given in Table 7.6 might be represented in economic terms (monetary units); for example, for X_6 (maximize economic benefits to local community), X_7 (maximize utility company profits), X_8 (improve balance of payments), and X_9 (reduce dependency on foreign fuels) the monetary units might be expressed as

$$\left. \begin{array}{l} X_{6_i} = \alpha(C_{DC} + C_O) \\ X_{7_i} = C_I - (X_8 + C_{DC} + C_O + C_{INS} + C_{DMS_i}) \\ X_{8_i} = \beta(X_{7_i}) - C_{DMS_i} \\ X_{9_i} = -C_{DN} \end{array} \right\} \quad (7.2)$$

where C_{DC} = design and construction costs;
C_O = operation and maintenance costs;
C_I = gross income from consumers;
C_{INS} = company insurance premiums;
C_{DMUi} = damage costs caused by the occurrence of some states of nature – loss of life, injury, environmental and physical damage, loss of production, court costs, punitive damages;

α = proportion of design, construction, operation and maintenance costs spent in the local community;
β = tax rate on company profits;
C_{DMSi} = damage costs borne by society (e.g. insurance payouts, government compensation); and
C_{DN} = costs of doing nothing (cost of importing products, power restrictions).

In the present case, the expected value for these economic attributes can then be expressed as

$$EV = \sum_{i=1}^{M} p_i (X_{6i} + X_{7i} + X_{8i} + X_{9i}) \qquad (7.3)$$

If system risk is defined as (probability of occurrence) × (consequences), then it is clear from Equation (7.1) that the expected value has the same measure as system risk. For example, both expressions may be given in terms of expected dollar losses per year (or expected fatalities per year). Thus, in some cases, expected values may also be referred to as system risks.

A risk-cost-benefit analysis is a straightforward extension of cost-benefit analysis, in which expected values are used (based on probabilistic expectation). In such an analysis, the consequences (benefits and losses) are generally associated with the 'normal' operation of the system being considered. However, exposure to continuous emissions may result in longer term or latent health problems to the workforce or to the local population: these consequences generally are difficult to quantify. Similar comments apply to catastrophic system failure, resulting in property and environmental damage and injuries and loss of life. The magnitude of these losses are more difficult to quantify owing to the various uncertainties surrounding estimates. Nevertheless, Meyer (1984) has estimated that the consequences of core melt-down at a nuclear power plant could cause property damage of approximately $14 billion and over 48 000 deaths over a 30-year period (early and latent fatalities). The question which arises is how these two estimates can be brought to a common value system.

There is no clear agreement on the economic value of a human life. Several methods have been proposed; these include:

1. 'human capital approach' (forgone earnings due to premature death – $450 000) (Marin, 1992);
2. 'value of a statistical life' (equal to $D.x where $D is the amount an individual is prepared to pay to reduce their fatality risk by 1^{-x}, hence group of x people would average one death less per year – $1.6 to $8.5 million) (Fischer et al., 1989);

3. money spent on government programmes per life saved ($100 000 for steering column protection to $90 million for asbestos removal) (Lind et al., 1991); or
4. government compensation payable for death by accident.

An alternative measure may be 'life-years' saved or lost (Lind et al., 1991). Using these figures, the economic loss of 48 000 deaths would range from $5–$400 billion. Finally, the method selected to estimate a monetary value on a human life may raise moral issues (e.g. Marin, 1992).

As if the difficulties in evaluating human life were not enough, the evaluation of the attributes associated with environmental and social values create even more problems. A typical measure of environmental and social values is 'quality of life'; defined by Power (1980) as the influences of the physical and social environment on human well-being (e.g. injury, clean air). However, estimating the economic value of quality of life is unclear because this is influenced by such present and future considerations as life expectancy, wealth (or resources) of the society and other demographic considerations.

It should be evident that the attribute value chosen (estimated) for a high consequence 'system failure' may differ by several orders of magnitude (e.g. $5 billion – $400 billion for 48 000 deaths). This will be reflected directly in the outcome of the decision analysis process, as it will usually be sensitive to the selected values of the attributes. For this reason, decision analyses outcomes should be subject to sensitivity analyses to ensure that decisions are not unduly influenced by estimates of consequences (see section 6.6). This is particularly the case for low probability/high consequence systems.

Losses from catastrophic system failure cannot always be remunerated from available corporate assets, insurance cover or even government resources. It is also likely that engineering systems operating in the 'low probability/high consequence' (LP/HC) arena would not be able to continue to operate if the designers, owners and operators of the engineering system were to bear all costs associated with system failure should it occur. Accordingly, some governments have enacted limitations on liability for some engineering systems (often partly in self-protection). For example, (1) in the United States the liability of nuclear licensees for third-party property damage and personal injury resulting from an accident at a nuclear power plant is limited to $7 billion by the Price-Anderson Act (Johnson, 1990) and (2) in Germany the liability for physical injury caused by defective pharmaceutical products is limited to $80 million (Jasanoff, 1984). Given that up to $400 billion in damages could occur if core melt occurred at an existing nuclear power plant in the US, it is readily apparent that these liability limits would not adequately compensate the victims of catastrophic system failure. Hence a significant proportion of expected losses, in the event of catastrophic system failure, will be borne by the victims and by society as a whole.

The influence of time is an important consideration since attribute values seldom occur simultaneously. Where future consequences can be represented in monetary terms, they may be discounted to equivalent current consequences, using the discount rate for interest-bearing investments (noting that the choice of discount rate is subject to considerable uncertainty). The time period over which consequences are discounted will also influence the decision, even though the period to be used is seldom well defined. For example, decisions relating to nuclear waste repository sites examine long-term consequences over a 'hundred thousand years', while consequences related to siting of a new airport might extend to only 30 years.

A further use of expected values is the computation of the 'cost-effectiveness' of various system configurations (e.g. Keeney and Raiffa, 1976). Such estimates are generally measured in terms of costs of risk reduction (e.g. safety systems) needed to prevent one premature death. The estimate of cost-effectiveness is then compared to a specified acceptance criteria; for regulatory authorities the acceptance criteria typically vary considerably, from approximately $1 million (Office of Management and Budget) to $5 million (USNRC) per life saved (e.g. Philley, 1992).

The attributes used to calculate the expected value often will represent objectives of more than one interested party, as illustrated in Table 7.6. Maximizing the expected value (Equation 7.3) will maximize the benefits only for society as a whole. Individuals or groups of interested parties might benefit rather more than others from the selected decision. Others might still receive minimal benefits or even suffer adverse consequences. Some of these problems may be overcome by adopting utility theory, as described in the next section.

7.4.3 Expected utility analysis

An expected value analysis provides no guidance for the combination of preferences if the objectives differ between the interested parties. For example, the siting of a nuclear power plant is a complex issue because of multiple conflicting objectives between the interested parties. In practice most decision-making scenarios are not simple, but are complex value problems. Utility theory has some merit because it attempts to consider and combine preferences of the interested parties across a range of economic, environmental and social values.

Utility theory provides a means of evaluating the risk preferences of the interested parties under choice uncertainty. A utility function $u_{X_n}(x_n)$ for each attribute X_n is used to represent these preferences. For example, if the decision-maker prefers outcomes A to B, B to C, and A to C; then the utility function must be formulated such that

$$u(A) > u(B) > u(C) \qquad (7.4)$$

Two arbitrary values may be assigned to the most and least preferable outcomes; for example, u(A) = 1, u(C) = 0 or u(A) = 100, u(C) = −50. It then follows that the value of u(B) will fall somewhere between u(A) and u(C); for example, the value of u(B) may be expressed as

$$u(B) = p \times u(A) + (1-p)\, u(C) \tag{7.5}$$

where p is selected such that the decision-maker is indifferent (i.e. outcomes are equally preferable) between selecting consequence B and a lottery in which he or she would receive consequence A with a probability p and receive consequence C with a probability of (1−p). This process may be repeated for other expected outcomes that fall between A and C. This establishes the 'utility function', in this case as a stepwise function. If the number of expected outcomes is large then the utility function might best be represented by a continuous function. It should be recognized that determining the value of p for different outcomes is subject to uncertainty and that the preferences will vary from decision-maker to decision-maker.

Utility functions may exhibit 'risk averse', 'risk neutral' or 'risk prone' attitudes. Also, the function may be 'monotonic' or 'non-monotonic'. To explain these terms it will be convenient in the following discussion to consider only economic attributes; however, the following concepts are equally valid for non-monetary attributes. A 'risk neutral' or linear utility function (see Figure 7.2) implies that a decision will be made solely on the expected monetary value. This may be accurate for decision-making in governments or large companies that can afford to sustain a loss (x″) against a 50–50 chance (i.e. p = 0.5) of making a substantial profit (x′). However, this is unlikely to be the case for individuals making decisions involving monetary values that are large in relation to the money at

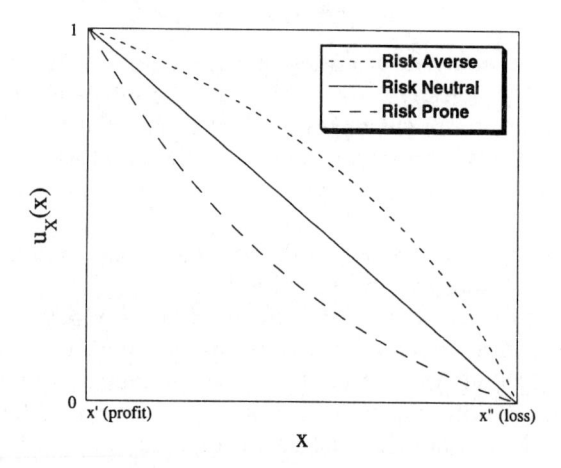

Figure 7.2 Monotonically decreasing utility functions.

their disposal (i.e. their working capital) (Benjamin and Cornell, 1970). Therefore, it is likely that these decision-makers would take a gamble only if the risk of loss (1-p) is small (e.g. p > 0.8, say). For larger risks (e.g. p < 0.8, say) these decision-makers might prefer to take no risk at all (i.e. to avoid the gamble) and to settle for the 'guaranteed' expected outcome (i.e. the average value of the expected profit and the expected loss, or, (x"+ x')/2). Such a decision-maker is 'risk averse'. However, once a decision-maker has suffered some large monetary loss (e.g. bankruptcy), he or she might well be numbed to further losses and might be willing to take quite large risks (e.g. p << 0.8). Such a decision-maker is 'risk prone' (Markowitz, 1952).

The utility functions in Figure 7.2 all show decreasing utility with an increase in the attribute scale. Alternatively, utility functions may increase as the attribute scale (e.g. profits) increases. A 'non-monotonic' utility function is characterized by both increases and decreases of the utility function. A situation where this may arise is the capacity of a proposed industrial plant; sufficient capacity to just meet demand is probably more desirable than lower (e.g. unmet demand – reduced profit) or higher (e.g. oversupply and wastage – reduced profit) capacities; see Figure 7.3. As an example, the transportation of hazardous chemicals has associated with it utility functions describing the consequences (which might result in fatalities, property and environmental damage), see Figure 7.4. In this case the severity of environmental damage might be described by a subjective index, such as that shown in Table 7.7.

As noted earlier, the objective of the decision-making process is to maximize expected utility. Since most decisions relate to complex value problems that involve group decision-making, the (supra) decision-maker needs to resolve a multiple objective problem. This may be done

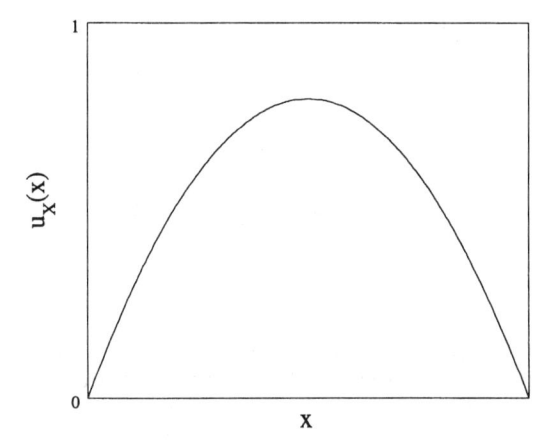

Figure 7.3 A non-monotonic utility function.

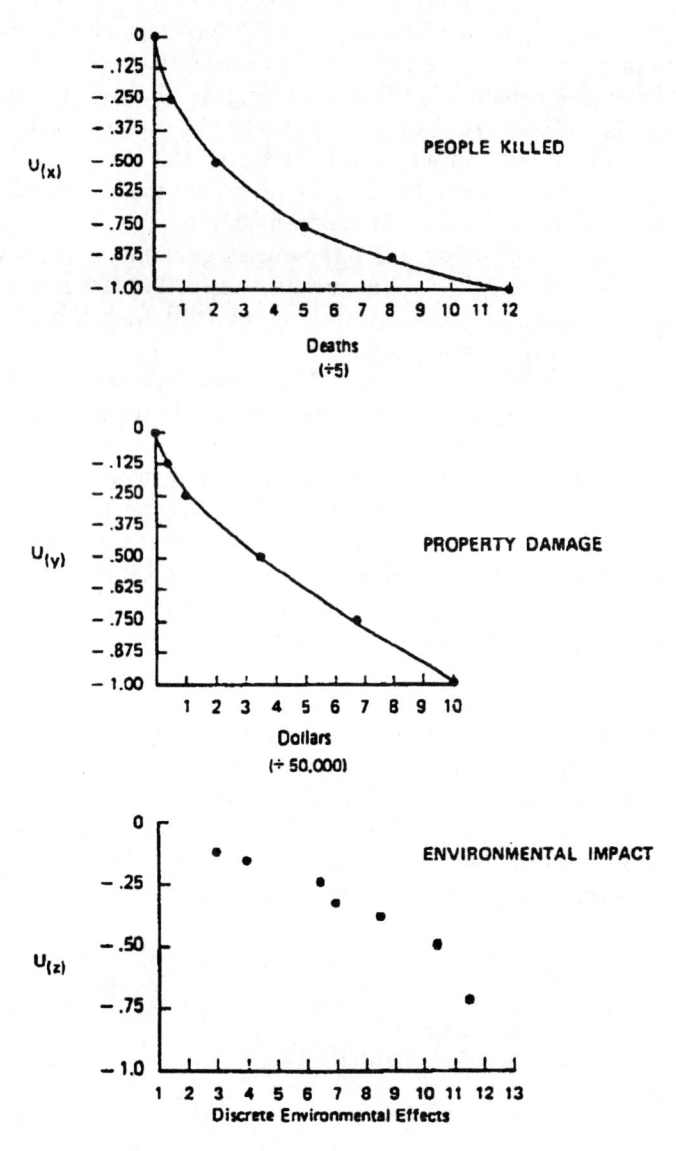

Figure 7.4 Utility functions for evaluating the transportation of hazardous chemicals.
Source: Kalelkar et al. (1974).

by considering the attributes (i.e. utility functions) of all the interested parties; these must be aggregated to produce a multi-attribute utility function. Several methods may be used to do this; each method is dependent upon attribute interdependencies. A typical multi-attribute utility function might be expressed as:

Table 7.7 Attribute measuring environmental damage of a hazardous chemical spill

Attribute value	Effect
1.	No effect.
2.	Residual surface accumulation of harmless material such as sugar or grain.
3.	Aesthetic pollution (odour-vapours).
4.	Residual surface accumulation of removable material such as oil.
5.	Persistent leaf damage (spotting, discolouration) but foliage remains edible for wildlife.
6.	Persistent leaf damage (loss of foliage) but new growth in the following year.
7.	Foliage remains poisonous to animals (indirect cause of some death upon ingestion).
8.	Animals become more susceptible to predators because of direct exposure to chemicals and resulting physical debilitation.
9.	Death to most smaller animals.
10.	Short-term (one season) loss of foliage with migration of specific animals that eat the foliage. Eventual reforestation.
11.	Death to foliage and migration of animals.
12.	Death to foliage and animals.
13.	Sterilization of total environment with no potential for reforestation or immigration of species.

Source: Kalelkar *et al.* (1974).

$$u_{X_1,X_2,\ldots,X_n}(x_1,x_2,\ldots,x_n) = \sum_{i=1}^{N} k_i u_{X_i}(x_i) \qquad (7.6)$$

where k_i is a scaling constant (between 0 and 1), $u_{X_i}(x_i)$ is the utility function for each attribute X_i and N is the number of attributes (utility functions). The scaling constants depend upon the preferences that the supra decision-maker has about each attribute; the size of each scalar constant is a measure of the influence of the attribute on the decision. Typical methods for the computation of scalar constants are described by Keeney and Raiffa (1976) and Goicoechea *et al.* (1982). Table 7.8 shows some attributes and their scaling constants obtained for a Mexico City airport study, where the supra decision-maker was the Mexican government. It is evident that there was more concern with the cost and capacity of the new airport than with the consequent noise levels. Presumably, the values of the scaling constants would change significantly if local residents were involved more closely in the decision-making process.

For n attributes the expected utility can be expressed as

$$EU_k[u_{X_1,X_2,\ldots,X_n}(x_1,x_2,\ldots,x_n)] = \sum_{i=1}^{M} p_i u_{X_1,X_2,\ldots,X_n}(x_1,x_2,\ldots,x_n) \qquad (7.7)$$

Table 7.8 Scaling constants for a Mexico City Airport study

Attribute	Attribute value	Scaling constant k_i
X_1 = total cost	millions of pesos	0.48
X_2 = capacity	operations/hr	0.60
X_3 = access time	minutes	0.10
X_4 = safety	number of people killed per aircraft accident	0.35
X_5 = displacement	number of people displaced by airport development	0.18
X_6 = noise	number of people subject to high noise levels	0.18

Source: Keeney and Raiffa (1976).

where k is the alternative being considered, p_i represents the probability of occurrence for each 'state of nature' i (with the probability obtained, as before, from system risk analysis), M is the number of states of nature, $x_1, x_2, \ldots, x_n$ represent attribute values associated with the occurrence of each state of nature and $u_{x_1, x_2, \ldots, x_n}(x_1, x_2, \ldots, x_n)$ is the multi-attribute utility function. 'States of nature' will include catastrophic system failures (e.g. release of a toxic substance into the environment), minor system failures (e.g. release of toxic substance within a building) and the remaining scenarios, in particular 'safe' operation. The expected utility is sometimes referred to as the 'subjective expected utility' if the probabilities of occurrence for each state of nature (p_i) are obtained subjectively. In such cases, these subjective probabilities are usually estimated directly from expert opinion.

Time is an important factor that may influence utility. Psychological studies show that utility of a decision-maker will increase and decrease with the postponement of adverse and beneficial consequences respectively (e.g. Kozielecki, 1975). This is important when decision-makers consider the possible long-term consequences (e.g. radioactive wastes generated by nuclear facilities) of the decision. Preferences may also change over time; for example, today the public is more aware of and concerned about environmental issues than it was two decades ago. This implies that uncertainties will arise both over possible future consequences and over future preferences. It is also likely that some decision variables (e.g. X_i, p_i) cannot be described adequately by point estimates (see Chapter 4). In general it is more realistic to represent such variables as random variables, in which the variance is a (first-order) measure of the uncertainty of the variable.

Note that the expected utility given by Equations (7.6) and (7.7) is equal to the expected value given by Equation (7.1) if

1. all the attributes are given in similar terms (e.g. costs),
2. all utility functions are risk neutral ($u_{X_i}(x_i) = X_i$) and
3. all scalar constants are equal to unity.

Regulatory safety goals

7.4.4 Other techniques

Other decision-making techniques have been proposed for decisions with multiple objectives; see Larichev (1983) and Goicoechea et al. (1982) for further details. However, most methods (if used by the same decision-maker) will generally produce similar decisions. It appears that the expected value and expected utility methods are the only rational methods that have been validated in any substantial manner (Larichev, 1983).

7.5 REGULATORY SAFETY GOALS

7.5.1 Types of regulation

Rather than use the results of a risk analysis as part of the decision-making process, based on economic or utility function arguments, the results can be used rather more directly for regulatory purposes. Typically, regulatory authorities are public institutions formed by government legislation. Their role is to develop general safety goals and set specific safety standards, to monitor system performance, and to prosecute individuals or companies if specified safety (or related) standards are violated. The safety standards used by regulatory authorities may be surrogates for the risks and hazards which are considered to be acceptable (or tolerable) to society.

In most cases, proponents or operators of engineering systems will need to obtain licences, permits or approvals (from one or more regulatory authorities) prior to the commencement of each stage of a project. Existing projects may need to undergo a similar approval process if modifications or expansions are proposed. In addition, existing systems may be subject to regular safety reviews by regulatory authorities to ensure continuing compliance with existing or revised regulatory safety standards.

Litigation based on breaches of common law (law of torts) may supplement regulatory requirements if the existing regulations do not adequately protect the well-being of individuals or if existing regulations are not appropriate for new or emerging technologies. The statute of limitations, difficulty in proving negligence (i.e. duty of care), high legal costs and delays associated with most private litigations tend to discourage many individuals from seeking legal redress. These problems may be partly overcome if a group action is initiated. Common law may also be used to obtain a court injunction in order to prohibit the development or operation of a particularly hazardous facility (Green, 1980). Moreover, criminal law may be used to prosecute individuals (e.g. operators, company directors) who have committed manslaughter or

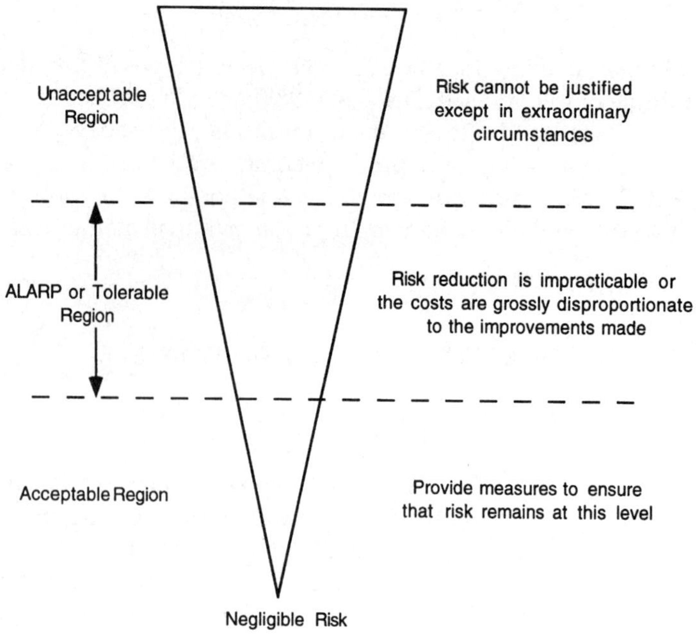
Unacceptable Region
Risk cannot be justified except in extraordinary circumstances
ALARP or Tolerable Region
Risk reduction is impracticable or the costs are grossly disproportionate to the improvements made
Acceptable Region
Provide measures to ensure that risk remains at this level
Negligible Risk

Figure 7.5 Levels of risk and ALARP.
Source: adapted from Sharp *et al.* (1993).

breached workplace legislation. However, juries are reluctant to convict individuals because it is often difficult to prove beyond reasonable doubt that gross negligence has occurred. For these and other reasons, Furmston (1992) concludes that common and criminal law often fail to provide an effective deterrent against unsafe behaviour. Thus, governments often resort to legislative action (e.g. creation of regulatory authorities) to protect its citizens against hazardous activities.

The risk acceptance criteria generally adopted by the US Nuclear Regulatory Commission, UK Nuclear Installation Inspectorate, UK Health and Safety Executive and other regulatory authorities is that risks and hazards should be 'As Low As Reasonably Possible' (ALARP) or 'As Low As Reasonably Attainable' (ALARA). The levels of risk and ALARP are shown in Figure 7.5. The definition for such terms as 'low', 'reasonably', 'possible' and 'attainable' are highly subjective and so attempts have been made to define this criteria in more tangible (and verifiable) terms – this often means in probabilistic terms. Hence, a risk-based approach may be used to provide a somewhat 'rational' basis for the development of safety goals.

Regulatory safety goals are increasingly being developed from risk-based ideas, including the concepts of perceived risks and formal

Regulatory safety goals

decision analysis discussed earlier. These may also be used to develop the following three types of regulatory safety goals:

1. specification standards (or deemed-to-comply);
2. quantifiable performance requirements; and
3. quantitative safety goals.

The latter two (quantifiable performance requirements and quantitative safety goals) are performance-based standards, that is, they specify the standard of performance (or safety objective) required to be achieved. It is then the responsibility of the licensee or operator to select the system configuration, operating practices and organizational structure by which the performance levels can be met. This allows flexibility of action and choice in seeking a solution which will comply with the regulation(s). Increasingly, it is the case that it is the responsibility of the licensee or operator to provide evidence of compliance – an important ingredient in the 'safety case' approach (see section 7.5.2).

Specification standards on the other hand are essentially a 'cookbook' type approach. Compliance to the set requirements will guarantee acceptance. Evidently, this approach does not provide much flexibility for the operator or licensee. It also tends to be adversarial and has the tendency to generate an atmosphere in which the operator or licensee may be tempted to see whether it is possible to 'get away with it'.

Other acceptance criteria used by regulatory authorities include the implementation of, and compliance to, quality assurance measures and standards (e.g. ISO9000 series; AS/NZS 4360–1995). Typical quality assurance measures include the requirement to conduct regular hazard scenario analyses (e.g. HAZOP, FMEA) and their findings being promptly addressed, development of written operating procedures, development of emergency procedures, the implementation of safety management systems, and auditing of safety management systems at regular intervals (e.g. Corbett, 1990; HSC, 1992).

It is important to recognize that regulatory safety goals are based mainly on past experience (Allen, 1992). Thus current regulatory safety goals may need to be revised when relevant new knowledge is available, such as that obtained from increased operating experience, new toxicological data, the occurrence of 'unexpected' system failures, or the development of new technologies.

7.5.2 Demonstrating compliance – the 'safety case'

For specification standards it is sufficient, usually, for the operator or licensee to allow inspection of the facility by regulatory personnel. Such inspection (together with any follow-up processes) will confirm whether standards have been met. As noted above, this may cause a degree of adversarial interaction. It does not necessarily create a commitment by management to understanding of risks and to risk (and consequence) control (see also Chapter 1).

In other cases it is becoming increasingly the case that operators and licensees must argue a 'safety case' – a document describing the risk assessment study and how the regulatory safety goals have been satisfied. For chemical process plants and offshore platforms the procedure is well established. For chemical process plants it has been termed also a 'safety audit' (e.g. Kletz, 1992). The documents arguing the case are reviewed or audited by the regulatory authority to ensure that the study deals properly with the facility under discussion (completeness requirement), that appropriate event probabilities and consequences have been considered, and that the results show compliance to the relevant regulations (e.g. Moss, 1990).

The advantage of the safety case approach is that the onus of proof is put back on to the licensee or operator. The argument has to be put that all reasonable steps have been taken to comply with the regulations and that where such regulations cannot be met, the steps taken will still result in a facility or system that is acceptably safe. Evidently, this allows rather greater flexibility in meeting safety standards, and it also allows the application of ALARP/ALARA principles (see above).

7.5.3 Specification-standard regulations

Specification-standard or deemed-to-comply regulations specify industry-specific safety measures that must be adopted. Typical specification-standard regulations include:

1. minimum concrete cover to reinforcing steel (for a reinforced concrete beam) of 50 mm to achieve a two-hour fire rating (e.g. AS3600, 1994);
2. minimum earth cover of 2.4m for uranium mill tailings (EPA, 1983);
3. minimum distance of Liquefied Petroleum Gas (LPG) storage tanks (for an automotive retail outlet) from residential properties (DOP, 1993);
4. 'best available technology' approach that specifies technologies that reduces the greatest sources of risk without causing widespread economic hardship to individuals or companies (e.g. Russell and Gruber, 1987);
5. 'defence-in-depth' approach that specifies that redundancies, parallel systems, safety devices (e.g. emergency pumps) and safety features (e.g. firewalls) must be designed into the system; and
6. other procedures that must be observed to ensure the 'safe' operation of a system.

These and other regulations are obtained mainly from past experience, although they may be derived from a generic risk analysis (e.g. DOP, 1993). Typically, the regulations are generic, in that they apply to all sites and system configurations for a particular process or industry and are not site specific. Thus, specification-standard regulations do not require a risk analysis to be carried out for approval to be given for a project, provided the specifications are met.

As noted, specification-standard regulations may be developed from a risk-based analysis. One example is that of the US Environmental Protection Agency (EPA) which in 1983 assessed measures to control the release of radon from uranium mill tailings; radon gas can cause cancer if inhaled. It was estimated that 600 excess cancer deaths would occur in the next 100 years if there were no control measures (i.e. uranium mill tailings exposed to the environment). However, a risk analysis found that 95% of the deaths would be avoided (i.e. 570 lives saved) if the tailings were covered by 2.4 m of earth, at a cost of $500 million. A cover of 4.8 m would further reduce the risk such that 99.5% of deaths would be avoided (i.e. 597 lives saved), but at a cost of $750 million. The EPA rejected the latter option because the estimated number of expected deaths were subject to considerable uncertainty since values of 570 and 597 are not significantly different, and hence the extra $250 million 'probably buys nothing at all'. The EPA regulation (EPA, 1983) thus specified that the minimum earth cover for uranium mill tailings be 2.4 m (Russell and Gruber, 1987).

7.5.4 Quantifiable performance requirements

Quantifiable performance requirements provide quantitative, but not necessarily probabilistic, criteria against which compliance to the requirements can be directly calculated or measured. Typical quantifiable performance requirements include:

1. the probable maximum flood (PMF) must not overtop a dam (e.g. Lave et al., 1990);
2. limits to plant personnel and public exposure to hazardous or toxic substances (e.g. 50mSv y^{-1} for ionizing radiation – HMSO, 1985);
3. structural system must support minimum design dead, live, wind, earthquake and snow loads used in building design (e.g. AS1170.1; 1989);
4. ship must retain buoyancy and stability if one or more watertight compartments are 'laid open to the sea' (Caldwell, 1992);
5. minimum endurance time for a temporary safe refuge (on an offshore platform) to be one hour (HSC, 1992); and
6. future generations should not have to take any action to protect themselves from the effects of high-level radioactive waste (IAEA, 1989).

A number of quantifiable performance requirements have been formulated from a risk-based approach. For example, structural engineering design codes specify that the resistance of the structural element (e.g. column) must be greater than the applied actions (e.g. axial force obtained from loads); design equations used for calculating the loads, actions and resistances are also provided. Numerous risk analyses were used by the developers of the design code to cover the range of possible design combinations. This was required to ensure that use of code specified design equations will result in acceptable 'nominal' probabilities of structural failure. The approach taken was to require the nominal probability of failure to (1) be relatively constant for a range of structural materials and loading conditions and (2) not exceed a 'target' value of approximately 1×10^{-4} and 1×10^{-5} failures per year. This process is termed 'calibration', see Melchers (1987) for a detailed review. Probabilities of failure are termed 'nominal' because they are calculated using relatively simple methods which do not necessarily accurately reflect reality, but are useful design models nevertheless. The safety targets used for this approach are derived from previously accepted structural engineering practice generally deemed to be satisfactory to the community. They are used simply to compare one project against another. No attempt is made to compare the 'nominal' (or 'notional') probabilities with regulatory or societal based tolerable risk levels. The whole process is one of relative comparison rather than comparison to absolutes. It follows that the nominal probabilities have no validity outside the framework in which they have been employed (Melchers, 1993). The procedure allows the use of conventional 'factors of safety' and of design equations for loads and resistances in practical design – the probabilistic basis being hidden behind apparently conventional rules (although the precise numerical values are fixed by calibration). There is no need for structural designers to conduct risk analyses as part of the practical design process.

7.5.5 Quantitative safety targets

Specification-standard regulations and quantifiable performance requirements are applied mainly to systems, sub-systems and system elements whose performance is not site specific. However, for many engineering systems the performance and safety of the system are site specific, with performance influenced by site location (is it close to population areas?), toxicity of produced or stored materials, equipment used, and other variables. In these cases it is necessary and often mandatory to conduct a risk analysis in order to provide quantitative evidence that the configuration of the engineering system complies with one or more regulatory quantitative safety targets.

Quantitative safety targets usually are in the form of risk-based acceptance criteria. They are based on:

1. quantitative measures of absolute minimum individual safety requirements,

2. de-minimus values below which risks are viewed as 'trivial' (Byrd and Lave, 1987) and
3. in-between regions of risk where formal decision analyses are required to provide economic or social justification of the risk.

This approach follows the ALARP and ALARA principles introduced in section 7.5.1 (see Figure 7.5). Quantitative safety targets may apply to specific failure of systems or sub-systems (e.g. core melt, safety system failure), release of toxic materials, and plant personnel and off-site health effects. Ensuring that a risk satisfies a quantitative safety target does not in itself mean that the risk is acceptable. The quantitative safety target is a 'target' only and other non-probabilistic criteria are also important in judging the overall acceptability of risks.

Quantitative safety targets will vary within and between industries. This is not particularly surprising, since they often result from a complex decision-making process involving trade-offs and negotiation between interested parties. Further, it is likely that any new or proposed targets will be capable of being met by existing facilities (unless they are known to be unsafe), particularly if the economic and social implications of facility closure is likely to be unacceptable to governments and local communities. Consequently, quantitative safety targets for existing facilities tend to be less stringent.

7.5.6 Examples – quantitative safety targets

(a) Nuclear power plants in the United States

The US Nuclear Regulatory Commission (USNRC) 'Final Policy Statement on Safety Goals' includes two qualitative safety goals and two quantitative safety targets. The qualitative safety goals (based on the ALARA principle) are:

1. individual members of the public should be provided a level of protection from the consequences of nuclear power plant operation such that individuals bear no significant additional risk to life and health; and
2. societal risks to life and health from nuclear power plant operation should be comparable to or less than the risks of generating electricity by viable competing technologies and should not be a significant addition to other societal risks (USNRC, 1986).

These qualitative safety goals are quantitatively defined by the following quantitative safety targets (referred to by the USNRC as 'quantitative design objectives'):

1. the risk to an individual or to the population in the vicinity of a nuclear power plant site (i.e. 1 mile) of 'prompt fatalities' (i.e. premature

deaths) that might result from reactor accidents should not exceed 0.1% of the sum of prompt fatality risks resulting from other accidents to which members of the US population are generally exposed - 5×10^{-7} per year; and

2. the risk to an individual or to the population in the area near a nuclear power plant site (i.e. 10 miles) of cancer fatalities (i.e. latent) that might result from reactor accidents should not exceed 0.1% of the sum of cancer fatality risks resulting from all other sources - 2×10^{-6} per year.

In 1982 the population of the United States was 231 million. Approximately 95 000 and 440 000 people died as a result of accidents and cancer, representing fatality rates of approximately 5×10^{-4} and 2×10^{-3} per year respectively (Spangler, 1987). Using these figures and the above safety goals leads to quantitative safety targets of 5×10^{-7} per year (i.e. 0.1% of 5×10^{-4}) for prompt fatality risks in the population within 1 mile of the exclusion boundary. The corresponding latent fatality risks in the population within 10 miles of the power plant is 2×10^{-6} per year. Note that these quantitative safety targets are applied quite separately to those due to system risks for initiating internal events (equipment failure, operator error) and due to system risks for external events (fire, earthquake) (USNRC, 1989).

(b) Nuclear power plants in the United Kingdom

For nuclear power plants in the UK the primary quantitative safety targets are the responsibility of the electricity generating authorities. The Nuclear Installation Inspectorate (NII) must accept these selected quantitative safety targets, but has responsibility for reviewing the risk analyses conducted by the electricity generating authorities. The primary quantitative safety targets are:

1. all accidents leading to a small release of radioactivity to be less than 1×10^{-4} per year;
2. individual accident sequences leading to a large uncontrolled release (e.g. large-scale core melt) to be less than 1×10^{-7} per year; and
3. all accident sequences (collectively) leading to a large uncontrolled release to be less than 1×10^{-6} per year (Cannell, 1987).

(c) Potentially hazardous industries in Australia

In overall urban planning the siting of potentially hazardous industrial developments is a critical issue. One approach is to apply quantitative safety targets for use as risk acceptance criteria (DOP, 1992). In this application it is customary for the outcomes of a probabilistic risk assessment to be given in terms of risk contours (i.e. risk is dependent on the distance from the source of the hazard). It is observed from Table 7.9 that a

Table 7.9 Suggested quantitative safety targets for various land uses

Land use	Individual fatality risk ($\times 10^{-6}$ per year)
Hospitals, schools, child-care facilities, old age housing	0.5
Residential, hotels, motels, holiday resorts	1
Commercial developments including retail centres, offices and entertainment centres	5
Sporting complexes and active open space	10
Industrial	50

Source: DOP (1992).

quantitative safety target of 1×10^{-6} fatalities per year has been suggested for residents in the vicinity of an industrial development. Table 7.9 also shows that the quantitative safety targets are influenced by the vulnerability of the population and their ability to take evasive action in the event of a release of hazardous materials. For this reason, the quantitative safety targets for nearby hospitals and schools are more stringent than those for nearby commercial premises, and less stringent for the actual site of the industrial development since most workers are highly mobile and their risk is somewhat voluntary. Nonetheless, the guidelines suggest that additional safety updates, reviews and risk reduction programmes be implemented at sites with an individual fatality risk in excess of 1×10^{-6} per year.

(d) Potentially hazardous industries in the Netherlands

The following quantitative safety goals for new and potentially hazardous facilities have been adopted in the Netherlands (Ale, 1991):

1. maximum individual risk to the public of 10^{-6} per year (i.e. 1% of the lowest 'natural death' rate for the population);
2. individual risks to the public of less than 10^{-8} per year are considered negligible;
3. maximum level of societal risk is 10^{-5} per year for an incident with a maximum of 10 deaths – a consequence 'n' times greater will result in a maximum risk level 'n^2' smaller; and
4. societal risks of less than 10^{-7} per year for an incident with a maximum of 10 deaths are considered negligible.

The above criteria are, presumably, influenced by the high population density. Nevertheless, for existing facilities the criteria may be relaxed by an order of magnitude.

(e) Cancer risks in the United States

In a review of 132 US federal government regulatory decisions associated with public exposure to environmental carcinogens it was found that

1. regulatory action always occurred if individual lifetime risks exceeded 4×10^{-3} and 3×10^{-4} for exposures to small and large population groups respectively; and
2. regulatory action never occurred if individual lifetime risks were less than 1×10^{-4} and 1×10^{-6} for exposures to small and large population groups respectively (Travis et al., 1987).

This suggests that lifetime (cancer) risks in the range of 1×10^{-4} and 1×10^{-6} (annual fatality risk of 1.3×10^{-6} and 1.3×10^{-8}) are judged as acceptable to society, a result in broad agreement with that reported by Rodricks and Taylor (1989). In some cases regulatory authorities have accepted risks greater than given by (1) above if

(a) The number of exposed individuals is very low;
(b) 'best available technology' was already in use; and
(c) further regulations could have resulted in closure of specific facilities.

For risks greater than (1) and less than (2) it was observed that cost-effectiveness was the main reason for further regulation, it being noted that risks were only reduced if the cost of risk reduction was less than $2 million per life saved. The study found the regulatory authorities to be reasonably consistent in their quantitative interpretation of acceptable risk (Travis et al., 1987).

(f) Other engineering systems

For some other engineering systems typical quantitative safety targets have been given in the literature. These are, in brief:

1. 1×10^{-9} failures per hour for flying control systems in modern aircraft (Hood, et al., 1992);
2. approximately 10^{-2} per flight for space shuttles (Paté-Cornell and Fischbeck, 1990);
3. 1×10^{-5} per year for an explosion of an LPG transport ship in harbour (Process Engineering, 1991);
4. 1×10^{-6} fatalities per year for failure of large dams (Lafitte, 1993); and
5. 1×10^{-3} for breach (within one hour) of a temporary safe refuge on an offshore platform (HSC, 1992).

The above quantitative safety targets have all been related to individual fatality risks and risks of system failure. However, quantitative safety targets may be expressed in other terms such as dose or exposure limits to toxic substances, risk of injury, property or environmental damage, and societal risk. For example, Figure 7.6 shows quantitative safety targets for societal risks that have been suggested or adopted by some European countries. It is noted that the quantitative safety targets become more stringent as the number of fatalities increases. This relationship tends to reflect the perception of risk in society; namely, that systems capable of killing hundreds of people in a single failure event (i.e. a 'catastrophic' potential) are of more concern than systems that kill fewer people more regularly (e.g. car accidents) – see also section 7.3.2.

(g) 'Global' quantitative safety targets

In most cases regulatory authorities have legislative responsibility only for a specific system or a restricted group of them. This means that inconsistencies could occur between the quantitative safety targets proposed or required for different systems, although, as observed above, quantitative safety targets tend to be rather consistent. The reason for this is that essentially the same methodology is used by most, if not all, regulatory authorities. It follows, therefore that it is possible to derive 'global' or generic safety targets from various inputs.

In this context it has been proposed that in the absence of industry regulations, the following 'global' quantitative safety targets may be used as preliminary guidelines for judging the acceptability of risks (Spangler, 1987):

1. annual fatality risks that exceed 1×10^{-3} are generally deemed as significant, obvious or unacceptable; regulation is mandatory (as found in a 1980 US Supreme Court ruling, see Byrd and Lave; 1987);
2. annual fatality risks in the range 1×10^{-3} to 1×10^{-6} are generally acceptable if the benefits outweigh the risks (i.e. satisfies a formal decision analysis); and
3. annual fatality risks smaller than 1×10^{-6} are deemed as negligible; further regulation is not warranted ('de-minimis' concept from common law where the court does not concern itself with trivia, see Byrd and Lave; 1987).

A comparison of these risk acceptance criteria with known risks is shown in Figure 7.7. Global quantitative safety targets have been suggested also by Renshaw (1990), Paté-Cornell (1994), and many others. Further, it should be recognized that these quantitative safety targets are 'global' and application to a specific case requires a degree of care and flexibility to allow for the special circumstances which might apply.

7.5.7 Some issues

The precision of a system risk estimate is not necessarily very high. In previous chapters it was seen that there are limitations and uncertainties

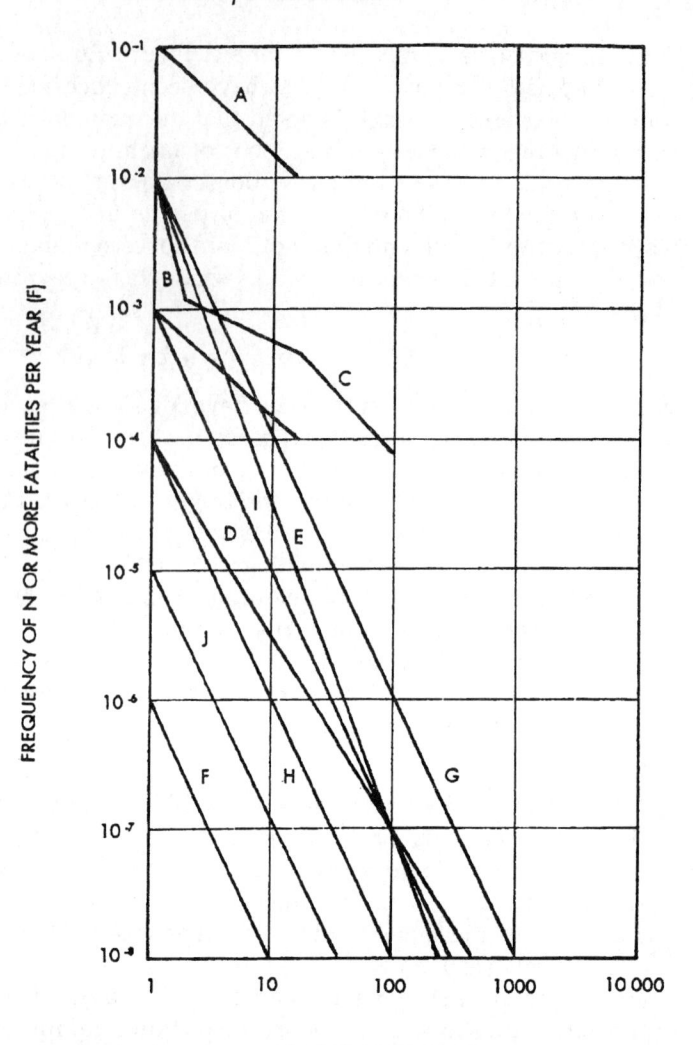

238   *Risk acceptance criteria*

[Figure: log-log plot with FREQUENCY OF N OR MORE FATALITIES PER YEAR (F) on y-axis ranging from 10^{-1} to 10^{-8}, and NUMBER OF FATALITIES (N) on x-axis ranging from 1 to 10 000, showing lines labeled A through J.]

A	Company 'X': unacceptable limit
B	Company 'X': 'no action' limit
C	Company 'Y': risk targets
D	UK nuclear industry: risk criteria
E	Groningen: unacceptable limit
F	Groningen: acceptable limit
G	Ale: unacceptable limit
H	Ale: acceptable limit
I	Netherlands: unacceptable limit
J	Netherlands: acceptable limit

Figure 7.6 Quantitative safety targets for societal risk.
Source: DOP (1992).

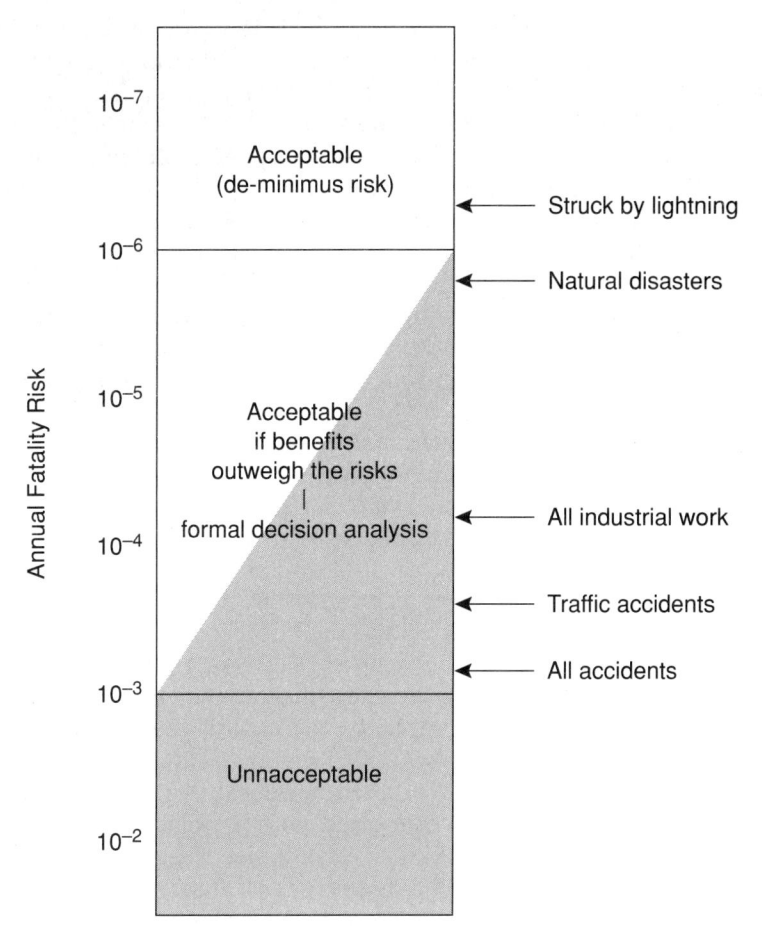

Figure 7.7 Quantitative risk acceptance criteria.
Source: existing risks obtained from Spangler (1987).

connected with system definition, system representation, standard of management systems, selection of models and parameter values, human error, and many other aspects of risk analysis procedures. It is to be expected, therefore, that there is a degree of imprecision in the estimation of system risk, as noted in Chapter 6. Moreover, a single-point estimate of system risk (obtained from, say, a QRA) provides no information about uncertainties (except through a sensitivity analysis). It is much more meaningful to undertake a PRA as this allows propagation of uncertainties through the system analysis. In this way the overall system risk may be represented by a probability distribution, with variance as a measure of system risk uncertainty (see section 6.3). However, the PRA approach can only propagate uncertainties if they are known.

Moreover, it cannot reflect uncertainties such as model selection, knowledge of the system and other uncertainties (or even errors), of which the analyst is unaware or which are incapable of being quantified. Moreover, it is not clear always whether quantitative safety targets should be compared with the mean, median or some upper confidence limit (e.g. 80%, 90% or 95% percentile) of system risk. To account for the probabilistic estimate of system risk, it has been suggested that 'prudent pessimism' should apply and that regulatory quantitative safety targets should refer to an upper confidence limit. This ensures that the risk acceptance criteria are used in a conservative manner. This is not universal; for example, it appears that the mean system risk is used by the USNRC in assessing compliance to quantitative safety targets (Paté-Cornell, 1994).

Given the imprecision of system risk estimates, it may be more appropriate to use system risk estimates only for 'comparative' or 'relative' risk purposes. This may include the prioritization of risk management measures (risk ranking) and 'calibration' with calculated system risks of similar projects (see section 7.5.4). In addition, as noted in Chapter 6, sensitivity analyses should be conducted to ascertain the effect of assumptions and uncertainties on the final decision.

The above discussion can be extended. Expressing quantitative safety targets as point estimates fails to reflect the variability and uncertainty as to what constitutes a risk acceptance criteria. It is probably more realistic to represent a quantitative safety target as a probability distribution, its shape representing the different technical, economic, social and political views of individuals concerned with establishing quantitative safety targets (Ra et al., 1993). Such a quantitative safety target can then be used to infer the probabilistic degree of acceptance of a system risk (e.g. 14% accept that risk), whereas point estimate quantitative safety targets simply denote either acceptability or unacceptability of risks.

If the system is subject to regulatory requirements imposed by more than one regulatory authority inconsistencies may arise. This may occur also if the regulations focus on only one aspect of system failure. An example is provided by Lave et al. (1990), who observed that the regulatory safety standards for dams were that they should be able to resist:

1. Probable Maximum Floods (PMF), having an annual likelihood of occurrence of 10^{-4} to 10^{-6};
2. earthquakes (having an annual likelihood of occurrence of 10^{-3}); and
3. wind (with an annual likelihood of occurrence of 10^{-3}).

Here the safety goal for dam failure through the occurrence of extreme floods is more stringent than for the other sources of dam failure, yet the potential consequences of dam failure are all identical.

Finally, there is general concern that the setting of quantitatively based regulatory safety goals may make the risk management process purely a 'numbers game' in which the prime objective becomes one of showing compliance, rather than one which is to assess, review and improve safety. This concern has some validity. It is therefore important that

1. quantitative safety targets must be accompanied by guidelines on models, reliability data, assumptions and methodology to be used in conducting risk analyses and risk assessments; and
2. quality assurance and peer review be undertaken to verify the quality of the risk analyses and risk assessments.

The objective of (1) is to ensure, in part, that system risks derived by different analysts will at least be comparable for similar system configurations. The second matter is discussed in the next section. As should be evident, many of the issues identified in this section are as yet unresolved. Readers might refer to Tweeddale (1993), Melchers (1993) and Paté-Cornell (1994) for more detailed discussions.

7.6 QUALITY ASSURANCE AND PEER REVIEW

Since the outcomes of risk analyses can affect significantly the design of facilities and their management, as well as regulatory and licensing decisions, it is important that some form of control be exercised. Usually this consists of quality assurance measures and peer reviews, the latter particularly for large systems. Quality assurance procedures tend to be focused on the review of internal procedures and practices. Peer review consists of an independent and critical review, conducted by recognized experts (in this case) in the area of risk analysis and risk assessment procedures.

Potential problem areas associated with the organization, technical work and documentation of a risk analysis are shown in Table 7.10. These matters may impact significantly on the quality of the risk analysis and its outcomes. In 12 peer reviews of nuclear power plant PRAs a number of significant issues were raised (see Figure 7.8); these include:

1. accident sequence and system modelling,
2. human performance analysis and
3. the identification of initiators.

Quality assurance measures should address these important issues. Guidelines described in EPRI (1983) suggest that normally this is accomplished by internal intradisciplinary reviews, in which work within a task is reviewed by other members working on the same task. Plant and design personnel should be included in the review team. In addition, an internal interdisciplinary review (i.e. one which extends beyond individual task boundaries) should be conducted to complement the intradisciplinary review.

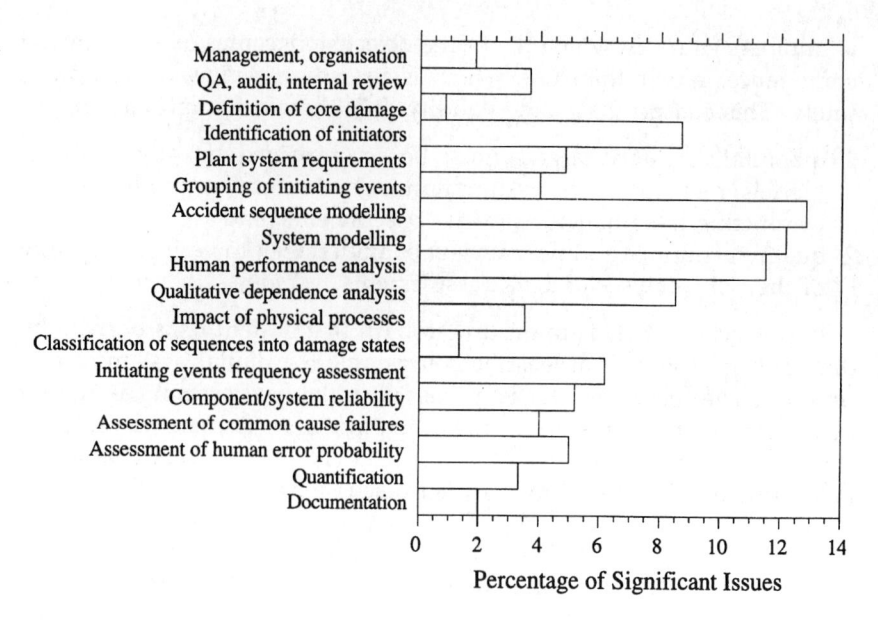

Figure 7.8 Significant issues in nuclear power plant PRAs.
Source: adapted from Gubler (1995).

Table 7.10 Potential problem areas that affect the technical adequacy of a PRA study

Area	Attribute
Organisation	Experience
	Balance
	Integration
	Communication
	Responsibility
Technical work	Completeness
	Accuracy
	Quantification
	Verification
	Consistency
Documentation	Clarity
	Traceability

Source: EPRI (1983). Copyright © 1983. Electric Power Research Institute. EPRI NP-3298. *An Approach to the Assurance of Technical Adequacy in Probabilistic Risk Assessments of Light Water Reactors.* Reprinted with permission.

Very comprehensive, generic guidelines for peer reviews for risk analyses of nuclear facilities, for example, have been provided by the International Atomic Energy Agency (IAEA, 1995). These guidelines state that the objectives of peer review are:

1. assess the adequacy of the treatment of important technological and methodological issues in the PRA (or risk analysis); and
2. assess whether specific conclusions and applications of the PRA are adequately supported by the underlying technical analysis (i.e. risk assessment).

It follows that a peer review also may satisfy quality assurance accreditation requirements. It is likely also to reduce organizational biases, analysis biases and will augment the experience of the analyst or regulatory authority. Moreover, Gubler (1995) remarks that 'an important aspect for a peer review is the communication and exchange of views between the international experts carrying out the review and the members of the PRA team'. Hence, a peer review will add to the quality of the risk analysis/assessment and the decision-making process and so enhance the credibility of decisions. This is particularly important for decisions which have significant potential political and public implications. Note that a peer review should complement, and not replace, regular internal reviews and other quality assurance measures.

7.7 SUMMARY

The acceptability of risks and hazards is influenced both by the perception of the risk to those likely to be exposed to it and the outcome(s) of formal decision-making processes. Regulatory safety goals tend to reflect both of these matters. The perceptions about a given level of risk are influenced by a range of psychological and socio-demographic variables which reflect the values, beliefs and attitudes of society, and in particular that part of society which is most likely to be exposed to the risk. Formal decision analyses such as expected value (risk-cost-benefit) and expected utility methods provide decision-makers with analytical techniques that assess preferences for possible consequences or outcomes. The acceptability of risks and the decisions made about them will vary over time, due to changes in social attitudes and priorities or the emerging knowledge of new hazards.

If properly enforced regulatory safety standards tends to reflect the level of risk and the degree of hazard which society is willing to tolerate. The safety standards may be set in a variety of ways: the most logical being those derived from a risk-based approach in which a risk analysis is needed to demonstrate compliance with the regulatory safety standards. Regulatory safety standards or goals for nuclear, chemical process and offshore industries specify that the risk of system failure must satisfy a specified set of quantitative safety targets. Because of a degree of imprecision with which risks can be assessed, the issue of the setting of risk acceptance criteria is controversial. Nevertheless, quantitative safety targets generally have been set in the range of 1×10^{-3} to 1×10^{-7} fatalities per person per year depending on the circumstances. Other measures such as risk of serious accident, risk of toxic release and risk of sub-system failure, are also used.

Although compliance with quantitative safety targets is an important criterion for risk acceptance, it should not be viewed as the sole criterion for decision-making. Other matters such as the successful implementation of other quantifiable performance requirements, quality assurance standards and economic and political issues generally will need to be considered as well.

Finally, Cox and Ricci (1989) conclude that 'acceptability of a technological risk is not only a matter of risk statistics and objective numbers, but of social processes and of trade-offs that society is willing to make to achieve the overall goal of decisions that are on average reasonably fair, efficient, workable, and acceptable'. It is also apparent that the issues associated with risk acceptance criteria are complex and subject to considerable uncertainty; many are not as yet completely resolved. For further reading on this topic, the reader may care to consult Philley (1992), Pidgeon et al. (1992), Hood et al. (1992), Reid (1992), Russell and Gruber (1987), Ricci et al. (1989), Wilson and Crouch (1987), Whipple (1987), Cox and Ricci (1989), Kemp (1991) and others.

REFERENCES

Ale, B.J.M. (1991) Risk analysis and risk policy in the Netherlands and the EEC, *Journal of Loss Prevention in the Process Industries*, **4**, 58–64.

Allen, D.E. (1992) The role of regulations and codes. In D. Blockley (ed.), *Engineering Safety*, McGraw-Hill, London, 371–84.

AS1170.1 (1989) *Minimum Design Loads on Structures: Part 1 – Dead and Live Loads and Load Combinations*, Standards Association of Australia, Sydney.

AS3600 (1994) *Concrete Structures*, Standards Association of Australia, Sydney.

AS/NZS 4360 (1995) *Risk Management*, Standards Association of Australia, Sydney.

Atallah, S. (1980) Assessing and managing industrial risk, *Chemical Engineering*, 8 September, 94–103.

Benjamin, J.R. and Cornell, C.A. (1970) *Probability, Statistics, and Decision for Civil Engineers*, McGraw-Hill, New York.

Brehmer, B. (1994) Psychology of risk characterisation. In B. Brehmer and N.E. Sahlen (eds), *Future Risks and Risk Management*.

Byrd, D. and Lave, L. (1987) Significant risk is not the antonym of de minimis risk. In C. Whipple (ed.), *De Minimis Risk*, Plenum Press, New York, 41–60.

Caldwell, J.B. (1992) Marine structures. In D. Blockley (ed.), *Engineering Safety*, McGraw-Hill, London, 224–67.

References

Cannell, W. (1987) Probabilistic reliability analysis, quantitative safety goals, and nuclear licensing in the United Kingdom, *Risk Analysis*, **7**(3), 311–19.
Corbett, R.A. (1990) Proposed OSHA safety regs target process plant procedures, *Oil and Gas Journal*, **88**(34), 80–4.
Cox, L.A. and Ricci, P.F. (1989) Legal and philosophical aspects of risk analysis. In D.J. Paustenbach (ed.), *The Risk Assessment of Environmental and Human Health Hazards: A Textbook of Case Studies*, Wiley, New York, 1017–63.
Cox, D., Crossland, B., Darby, S.C. et al. (1992) *Estimation of Risks From Observations on Humans, Risk: Analysis, Perception and Management*, The Royal Society, London, 67–87.
Delbridge, P. (1990) Public participation in risk assessments, *Plant/Operations Progress*, **9**(3), 183–5.
DOE (1990) *The Public Enquiry into the Piper Alpha Disaster*, Department of the Energy, Cm 1310, HMSO, London.
DOP (1992) *Risk Criteria for Land Use Safety Planning*, Hazardous Industry Planning Advisory Paper No. 4, Department of Planning, Sydney, Australia.
DOP (1993) *Liquefied Petroleum Gas Automotive Retail Outlets*, Hazardous Industry Locational Guidelines No. 1, Department of Planning, Sydney, Australia.
Douglas, M. (1987) *How Institutions Think*, Routledge, London.
EPA (1983) *Environmental Standards for Uranium and Thorium Mill Tailings at Licensed Commercial Processing Sites; Final Rule*, US Environmental Protection Agency, Federal Register, Vol. 48, 45926.
EPRI (1983) *An Approach to the Assurance of Technical Adequacy in Probabilistic Risk Assessment of Light Water Reactors*, Rep. EPRI-NP-3298, Electric Power Research Institute, Palo Alto, California.
Fischer, A., Chestnut, L.G. and Violette, D.M. (1989) The value of reducing risks of death: a note of new evidence, *Journal of Policy Analysis and Management*, **8**(1), 88–100.
Fischhoff, B., Lichtenstein, S., Slovic, P. et al. (1981) *Acceptable Risk*, Cambridge University Press, Cambridge.
Fischhoff, B. (1989) Risk: a guide to controversy. In *Improving Risk Communication*, National Academy Press, Washington, DC.
Furmston, M.P. (1992) Reliability and the law. In D. Blockley (ed.), *Engineering Safety*, McGraw-Hill, London, 385–401.
Gardener, C.T. and Gould, L.C. (1989) Public perceptions of the risks and benefits of technology, *Risk Analysis*, **9**, 225–42.
Goicoechea, A., Hansen, D.R. and Duckstein, L. (1982) *Multiobjective Decision Analysis with Engineering and Business Applications*, Wiley, New York.
Green, C.H. and Brown, R.A. (1978) Counting lives, *Journal of Occupational Accidents*, **2**, 55–70.
Green, H.P. (1980) The role of law in determining acceptability of risk. In R.C. Schwing and W.A. Albers (eds), *Societal Risk Assessment: How Safe is Safe Enough?*, Plenum Press, New York, 255–66.
Gubler, R. (1995) International Peer Review Service (IPERS) of the IAEA. In *Status and Experience*, PSA'95 Conference.
Gutmanis, I. and Jaksch, J.A. (1984) High-consequence analysis, evaluation, and application of select criteria. In R.A. Waller and V.T. Covello (eds), *Low-Probability High-Consequence Risk Analysis*, Plenum Press, New York, 393–424.
HMSO (1985) *The Ionising Radiations Regulations 1985*, Statutory Instrument 1985, No. 1333, HMSO, London.
Hood, C.C., Jones, D.K.C., Pidgeon, N.F. et al.(1992) Risk management, In *Risk: Analysis, Perception and Management*, The Royal Society, London, 135–92.

HSC (1992) *Draft Offshore Installations (Safety Case) Regulations 199-*, Health and Safety Commission, UK.
HSE (1988) *The Tolerability of Risk from Nuclear Power Stations*, Health and Safety Executive, HMSO, London.
IAEA (1989) *Safety Principles and Technical Criteria for the Underground Disposal of High-Level Radioactive Wastes*, International Atomic Energy Agency, IAEA Safety Series No. 99, Vienna.
IAEA (1995) *IPERS Guidelines for the International Peer Review Service*, International Atomic Energy Agency, IAEA-TECDOC-832, Vienna.
ISO 9000 (1987) *Quality Management and Quality Assurance Standards*, International Organization for Standardization.
Jasanoff, S. (1984) Compensation issues related to LP/HC events: the case of toxic chemicals. In R.A. Waller and V.T. Covello (eds), *Low-Probability High-Consequence Risk Analysis*, Plenum Press, New York, pp. 361–371.
Jasanoff, S. (1993) Bridging the two cultures of risk analysis, *Risk Analysis*, **13**(2), 123–9.
Johnson, J.W. (1990) Nuclear power and the Price-Anderson Act: an overview of a policy in transition, *Journal of Political History*, **2**(2), 213–32.
Kalelkar, A.S., Partridge, L.J. and Brooks, R.E. (1974) Decision analysis in hazardous material transportation, *Proceedings of the 1974 National Conference on Control of Hazardous Material Spills*, American Institute of Chemical Engineers, San Francisco, 336–44.
Keeney, R.L. and Raiffa, H. (1976) *Decisions with Multiple Objectives: Preferences and Value Tradeoffs*, Wiley, New York.
Kemp, R.V. (1991) Risk tolerance and safety management, *Reliability Engineering and System Safety*, **31**, 345–53.
Kletz, T.A. (1992) Process industrial safety. In D. Blockley (ed.), *Engineering Safety*, McGraw-Hill, London, 347–68.
Kozielecki, J. (1975) *Psychological Decision Theory*, D. Reidel Publishing Company, Dordrecht, Netherlands.
Lafitte, R. (1993) Probabilistic risk analysis of large dams: its value and limits, *Water Power and Dam Construction*, **45**(3), 13–16.
Larichev, O.I. (1983) Systems analysis and decision-making. In P.Humphreys, O.Svenson and A.Vari (eds), *Analysing and Aiding Decision Processes*, North-Holland, 125–44.
Lave, L.B. (1987) Health and safety risk analyses: information about better decisions, *Science*, **236**, 291–5.
Lave, L.B., Resendiz-Carrillo, D. and McMichael, F.C. (1990) Safety goals for high-hazard safety goals, *Water Resources Research*, **26**(7), 1383–91.
Lichtenstein, S., Slovic, P., Fischhoff, B. *et al.* (1978) Judged frequency of lethal events, *Journal of Experimental Psychology (Human Learning and Memory)*, **4**, 551–78.
Lind, N.C., Nathwani, J.S. and Siddall, E. (1991) Management of risk in the public interest, *Canadian Journal of Civil Engineering*, **18**, 446–53.
Marin, A. (1992) Costs and benefits of risk reduction. In *Risk: Analysis, Perception and Management*, The Royal Society, London, 192–201.
Markowitz, H. (1952) The utility of wealth, *Journal of Political Economy*, **60**(2), 151–8.
Melchers, R.E. (1987) *Structural Reliability: Analysis and Prediction*, Ellis Horwood, Chichester, England.
Melchers, R.E. (1993) Society, tolerable risk and the ALARP principle. In R.E. Melchers and M.G. Stewart (eds), *Probabilistic Risk and Hazard Assessment*, Balkema, Netherlands, 243–52.
Meyer, M.B. (1984) Catastrophic loss risks: an economic and legal analysis, and

a model state statute.In R.A. Waller and V.T. Covello (eds), *Low-Probability High-Consequence Risk Analysis*, Plenum Press, New York, 337–60.
Morone, J.F. and Woodhouse, E.J. (1989) *The Demise of Nuclear Energy? Lessons from Democratic Control of Technology*, Yale University Press.
Moss, T.R. (1990) Auditing offshore safety risk assessments, *Journal of Petroleum Technology*, **42**(10), 1241–3.
Otway, H.J. and Wynne, B. (1989) Risk communication: paradigm and paradox, *Risk Analysis*, **9**, 141–5.
Otway, H.J. and von Winterfeldt, D. (1982) Beyond acceptable risk: on the social acceptability of technologies, *Policy Sciences*, **14**, 247–56.
Paté-Cornell, M.E. (1984) Aggregation of opinions and preferences in decision problems. In R.A. Waller and V.T. Covello (eds), *Low-Probability High-Consequence Risk Analysis*, Plenum Press, New York, 493–503.
Paté-Cornell, M.E. and Fischbeck, P.S. (1990) *Safety of the Thermal Protection System of the Space Shuttle Orbiter: Quantitative Analysis and Organizational Factors – Phase 1: Risk-Based Priority Scale and Preliminary Observations*, Report to the National Aeronautics and Space Administration.
Paté-Cornell (1994) Quantitative safety goals for risk management of industrial facilities, *Structural Safety*, **13**, 145–57.
Philley, J.O. (1992) Acceptable risk – an overview, *Plant/Operations Progress*, **11**(4), 218–23.
Pidgeon, N., Hood, C., Jones, D. et al. (1992) Risk perception. In *Risk: Analysis, Perception and Management*, The Royal Society, London, 89–134.
Power, T.M. (1980) *The Economic Value of the Quality of Life*, Westview Press, Boulder, Colorado.
Process Engineering (1991) Refining attitudes towards a risky business, *Process Engineering*, **72**(4), 43–4.
Ra, K-Y., Lee, B-W. and Chang, S-H. (1993) A probabilistic safety criterion for core melt frequency based on the distribution of the public's safety goals, *Nuclear Technology*, **101**, 149–58.
Reid, S.G. (1992) Acceptable risk. In D. Blockley (ed.), *Engineering Safety*, McGraw-Hill, London, 138–66.
Renshaw, F.M. (1990) A major accident prevention program, *Plant/Operations Progress*, **9**(3), 194–7.
Ricci, P.F., Cox, L.A., and Dwyer, J.P. (1989) Acceptable cancer risks: probabilities and beyond, *Journal of the Air Pollution and Control Association, JAPCA*, **39**(8), 1046–53.
Rodricks, J.V. and Taylor, M.R. (1989) Comparison of risk management in U.S. regulatory agencies, *Journal of Hazardous Materials*, **21**(3), 239–53.
Rohrmann, R. (1995) Technological risks – perception, evaluation and communication. In R.E. Melchers and M.G. Stewart (eds), *Integrated Risk Assessment: Current Practice and New Directions*, Balkema, Netherlands, 7–13.
Russell, M. and Gruber, M. (1987) Risk assessment in environmental policymaking, *Science*, **236**, 286–90.
Sharp, J.V., Kam, J.C. and Birkinshaw, M. (1993) Review of criteria for inspection and maintenance of North Sea structures. In *Proceedings 1993 OMAE, Vol. 2, Safety and Reliability*, 363–8.
Slovic, P., Fischhoff, B., and Lichtenstein, S. (1980) Facts and fears: understanding perceived risk. In R.C. Schwing and W.A. Albers (eds), *Societal Risk Assessment: How Safe is Safe Enough?*, Plenum Press, New York, 181–216.
Slovic, P. (1995) Perceived risk, trust and democracy, *Risk Management Quarterly*, US Department of Energy, **3**(2), 4–13.
Spangler, M.D. (1987) A summary perspective on NRC's implicit and explicit

use of De Minimis risk concepts in regulating for radiological protection in the nuclear fuel cycle. In C. Whipple (ed.), *De Minimis Risk*, Plenum Press, New York, 111–43.

Tann, R.V. (1992) Transportation of nuclear materials. In R.F. Cox (ed.), *Assessment and Control of Risks to the Environment and to People*, The Safety and Reliability Society, Manchester, UK, 15/1–15/6.

Travis, C.C., Richter, S.A., Crouch, E.A.C. *et al.* (1987) Cancer risk management: A review of 132 Federal Regulatory decisions, *Environmental Science and Technology*, **21(5)**, 415–20.

Tweeddale, H.M. (1993) Maximising the usefulness of risk assessment. In R.E. Melchers and M.G. Stewart (eds), *Probabilistic Risk and Hazard Assessment*, Balkema, Netherlands, 1–11.

USNRC (1975) *An Assessment of Accident Risks in U.S. Nuclear Power Plants*, United States Nuclear Regulatory Commission, WASH-1400, NUREG-75/014, Washington, DC.

USNRC (1986) *Safety Goals for the Operation of Nuclear Power Plants; Policy Statement*, US Nuclear Regulatory Commission, Federal Register, Vol. 51, 30028.

USNRC (1989) *Severe Accident Risks: An Assessment for Five Nuclear Power Plants*, NUREG-1150, US Nuclear Regulatory Commission, Washington, DC.

Whipple, C. (1987) *De Minimis Risk*, Plenum Press, New York.

Wilson, R. and Crouch, E.A.C. (1987) Risk assessment and comparisons: an introduction, *Science*, **236**, 267–70.

APPENDIX A

Applications

A.1 INTRODUCTION

The application of probabilistic risk and hazard assessments to typical engineering systems may help to illustrate the essential ideas presented in this book; namely, the calculation, usefulness, limitations and uncertainties of probabilistic measures of system risk for engineering systems. It will be observed also that there is a large degree of commonality between the approaches and the risk analysis/assessment techniques employed for different systems. This is illustrated below with applications relating to

1. a nuclear power plant,
2. a chemical storage facility,
3. the Space Shuttle orbiter,
4. structural tension member design rules and
5. a gravity dam.

A.2 NUCLEAR POWER PLANTS

A study was conducted by the US Nuclear Regulatory Commission (USNRC) to provide an assessment of severe accident risks for five nuclear power plants, each of a different design (USNRC, 1989). The study aimed to (1) identify specific plant design and operational characteristics that produce risk vulnerabilities and therefore initiate appropriate management programmes and (2) compare system risk with USNRC safety goals. Estimates of system risk were obtained for core damage, accident progression, radiation containment and offsite consequences (i.e. radiation release and risk to public safety). A Probabilistic Risk Analysis (PRA) utilizing event and fault tree system representation was used as the main operational tool (see Figure A.1). Event tree logic was used for the Human Reliability Analysis (HRA) component of the PRA.

The PRA incorporated the effect of

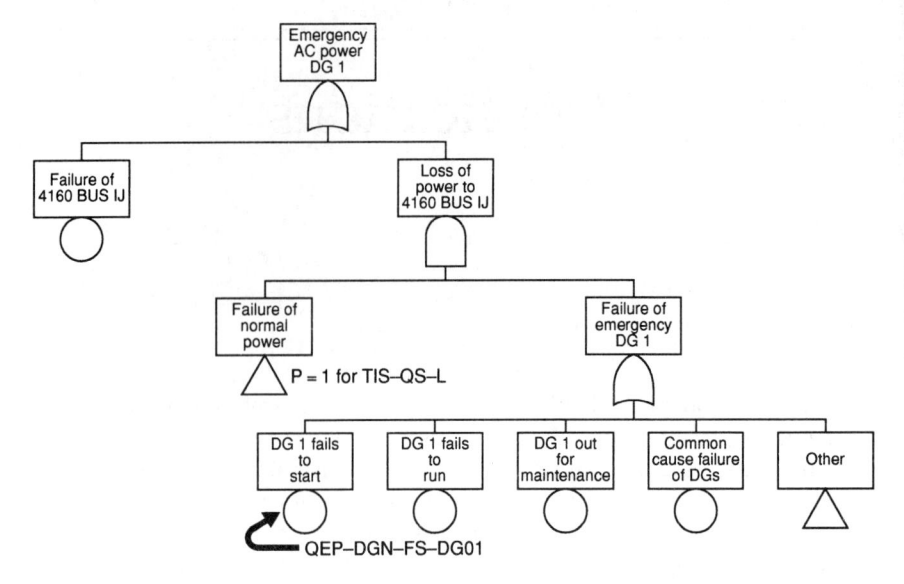

Figure A.1 Reduced fault tree for emergency diesel generator.
Source: USNRC (1989).

1. 'internal events':
    - Equipment failure
    - Operator error:
        (a) Pre-accident errors: Errors in normal operating conditions (e.g. incorrectly reading meter or turning dial) – mainly 'slip' type errors. Error events quantified by THERP database.
        (b) Post-accident errors: Failure to respond or recover from an accident condition – mainly 'mistake' type errors. Error events quantified using time-reliability correlations (e.g. HCR database).
2. 'External events':
    - Fire (e.g. in control room, auxiliary buildings)
    - Earthquake

Error rates and error factors for several tasks considered in the study are shown in Table A.1. The influence of air crashes, hurricanes and flooding were excluded from the analysis due to their low frequency of occurrence.

Uncertainties (or variabilities) in equipment and operator performance were propagated through the risk analysis, hence core damage frequency and other risks are represented as probability distributions, see Figure A.2. For the Surry Power Plant in Virginia (a pressurized water reactor of 788 MW capacity), the mean risks of core damage as obtained from the PRA are shown in Table A.2. The mean risks of core damage attributed to operator error were obtained from an analysis that incorporated the influence of human error. Equipment failure refers to an analysis in

Table A.1 Typical human error rates for nuclear power plant operations

Human error	Error rate	Error factor (EF)
Common-mode miscalibration of instrument	2.5E-5	10
Failure during isolation and repair of pump	3.0E-5	16
Operator fails to initiate level control	1.0E-3	5

Source: adapted from USNRC (1989).

Table A.2 Mean risks of core damage for Surry Nuclear Power Plant

		Mean risk of core damage (per year)
Internal events:	equipment failure	1.1E-5
	operator error	2.9E-5
External events:	seismic (LLNL)	1.2E-4
	seismic (EPRI)	2.5E-5
	fire	1.1E-5

Note: Seismic analyses used two data sources: LLNL (Lawrence Livermore National Laboratory) and EPRI (Electric Power Research Institute).
Source: adapted from USNRC (1989).

Table A.3 Public risk estimates (per year) for Surry Nuclear Power Plant

	Internal initiators	Fire
Early cancer fatality risk	2E-6	3E-8
Latent cancer fatality risk	5E-3	3E-4
Population dose within 50 miles of the site	5 rems	0.4 rems
Population dose within the entire site region	30 rems	2 rems
Individual early cancer fatality risk in the population within 1 mile of the NPP exclusion area boundary	2E-8	7E-10
Individual latent cancer fatality risk in the population within 10 miles of the NPP site	2E-9	1E-10

Source: adapted from USNRC (1989).

which all error rates for human error were equated to zero. Table A.2 shows that human error is a major factor contributing to risks of core damage.

Estimates of public risk (a measure of system risk) were obtained also from the PRA (see Table A.3). The USNRC has adopted the following quantitative safety targets for different measures of public risk:

1. 5×10^{-7} – average individual early fatality risk in the population within 1 mile of the nuclear power plant exclusion area boundary; and
2. 2×10^{-6} – average individual latent cancer fatality risk in the population within 10 miles of the nuclear power plant site.

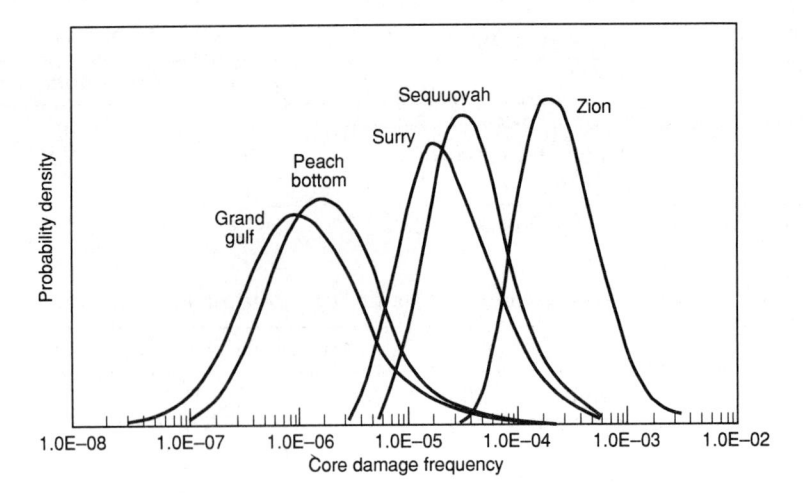

Figure A.2 Risk of core damage caused by internal events for five nuclear power plants.
Source: USNRC (1989).

These safety goals are based on the requirement that the measure of risk for NPP accidents should not exceed 0.1% of the sum of early or latent fatality risks resulting from all other causes to which members of the US population are generally exposed (Okrent, 1987).

Information on the consequences of off-site releases was incorporated into the PRA. It was found that the relevant estimates of public risk for internal initiating events were 'well below' the NRC safety goals for all five nuclear power plants, see Figure A.3. However, extensive and detailed recommendations were also made on appropriate design changes and management programmes to reduce further the risk and consequence of accidents.

A.3 CHEMICAL STORAGE FACILITY

A risk assessment was conducted by Boykin et al.(1984) to compare the risks associated with proposed improvements to a chemical storage facility in a large chemical process plant. The existing chemical storage facility had been in operation for 30 years; there had not been a major release of chemicals during this period. However, the storage of large quantities of chemicals was a major concern to the industry and to the public. The chemical storage system comprised three tanks, a circulating and cooling system, and an atmospheric venting system (see Figure A.4). The study considered two options for upgrading the system; namely,

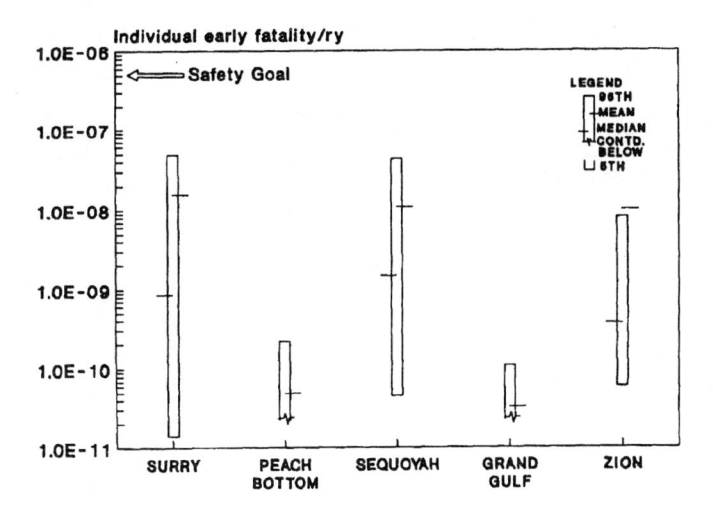

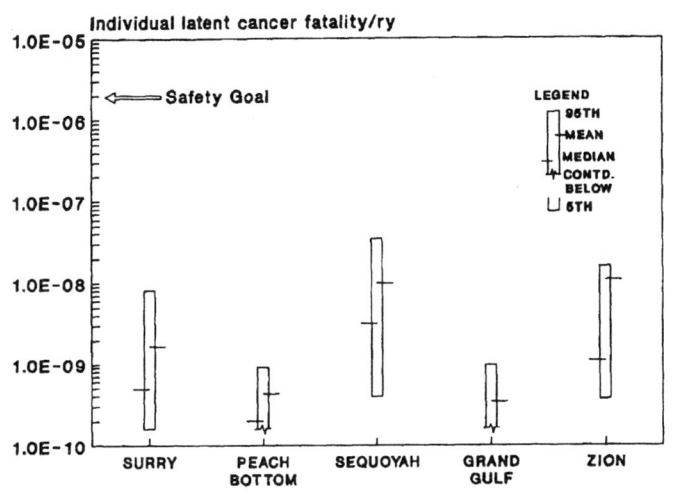

Figure A.3 Comparison of individual early and latent cancer fatality risks at all plants – internal initiators.
Source: USNRC (1989).

1. Proposed System 1: replace atmospheric venting system with a closed system safety venting system (cost – $12 million);
2. Proposed System 2: same as Proposed System 1 except with additional safety equipment (cost – $17 million).

Both of the proposed systems also include an upgrade of the cooling and ventilation systems.

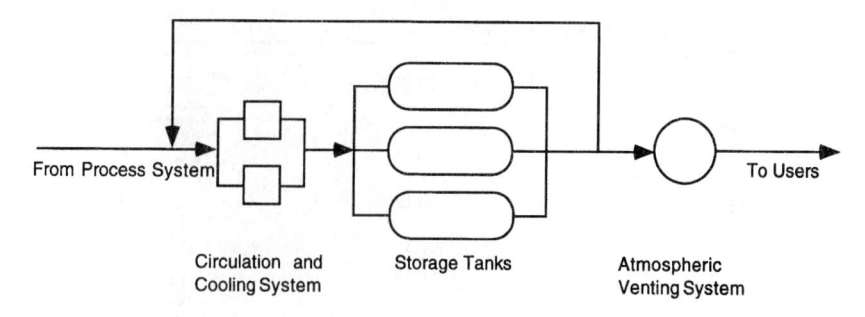

Figure A.4 A graphical representation of the chemical storage system. *Source:* adapted from Boykin et al. (1984).

System failure was considered in the present study in terms of an unacceptable level of the risk of death (to the public) due to a release of a toxic chemical from the storage area. Fault-trees were used to describe the sequence of events that could lead to system failure (over 40 pages of logic diagrams); a total of 140 accident events were identified. The quantification of equipment failure rates were obtained from (1) the Systems Reliability Services (SRS) data bank, and (2) plant maintenance records. Event-trees were then used to represent the influence of safety systems (e.g. flare stack, water deluge) and operator intervention in controlling any chemical release. Operator error rates were estimated using the 'paired comparisons' expert opinion procedure. System analysis of the fault-tree and the event-tree enabled estimates of the annual probability of a toxic release from the storage area to be calculated, see Table A.4.

Table A.4 Summary of annual probabilities of toxic release

Event	Annual probability
Current system:	
Uncontrolled release	3.6E-4
System leaks	5.2E-2
Fire	1.0E-4
Vessel breach	3.6E-8
Proposed system 1:	
Uncontrolled release	1.2E-7
System leaks	5.2E-2
Fire	1.0E-4
Vessel breach	1.2E-11
Proposed system 2:	
Uncontrolled release	2.3E-9
System leaks	5.2E-2
Fire	1.0E-4
Vessel breach	4.2E-10

Source: adapted from Boykin et al. (1984).

Chemical storage facility

Table A.5 Risk-cost comparisons for chemical storage facility

System	Risk of death per year[a]	Expected fatalities over plant life	Cost ($ millions)	Cost/life saved
Current	1.1E-6	2.2	0	–
Proposed 1	3.6E-10	0.0007	12	$5.5 million
Proposed 2	6.9E-12	0.00001	17	$7.7 million

Note: [a] Probability × consequence (20% fatality rate of exposed) divided by population at risk (65,000)
Source: adapted from Boykin et al. (1984).

Approximately 65 000 people live in an area where exposure to a released chemical cloud is possible. For a single accident, however, it was estimated that approximately 1000 people and 400 people would be exposed, for an uncontrolled release and for vessel breach respectively. There is no population exposure for a fire or system leak. It was then assumed that fatalities would amount to 20% of the exposed population. This allowed calculation of the risk of death per year (to the public), the expected fatalities over plant life (taken as 30 years) and the cost per life saved. The results are shown in Table A.5.

However, an additional analysis revealed that the process of installing the additional safety systems could pose a greater risk than having a chemical release without the proposed safety systems, see Table A.6. The risk increased because even a small chemical release (with no off-site consequences) can directly harm the construction workers. On the other hand, the study suggested that it may not be valid to make direct comparisons between voluntary risks to construction workers and involuntary risks to the public.

Using the risk-based approach presented herein as a basis for decision-making, management came to the following conclusions:

1. the risk of an accidental release in the current system was not acceptable due to the close proximity of the public to the plant (and the resulting concerns about large public liability exposure);

Table A.6 Construction risk comparisons

Event	Expected fatalities per year
Release:	
Proposed 1	2.4E-5
Proposed 2	4.6E-7
Safety system construction (per man-year)	5.0E-4

Source: adapted from Boykin et al. (1984).

2. the probability of an accidental release for the proposed systems was extremely small. However, the influence of risk to construction workers tended to negate these low estimates of societal risk.
3. it might be more 'cost-effective' to reduce the consequences of an accidental release. This could be achieved by constructing a new feed system that eliminates the need for a chemical storage system. Evidently, the amount of chemical that is then available for release would be reduced to such an extent that there was considered to be no likelihood of an off-site release. The new proposal could be implemented with a much lower capital cost than either of the other proposals.

It is of interest to note that the final alternative was not considered at the beginning of the study because of 'poor problem definition and the limited scope of the alternatives given to the analysts' (Boykin et al., 1984).

A.4 THERMAL PROTECTION SYSTEM OF THE SPACE SHUTTLE ORBITER

Paté-Cornell and Fischbeck (1990) have conducted a risk assessment of the Thermal Protection System (TPS) used on the Space Shuttle orbiter. The TPS comprises approximately 20 000 black tiles that shield the bottom of the orbiter during re-entry from temperatures of up to 2300 °F. The study developed a risk analysis model (essentially a HRA model since TPS performance is influenced mainly by human factors) in order to assess the probability of Loss of Vehicle (LOV) as a consequence of tile failure (i.e. the loss of one or more tiles). It was assumed that tile failure was due to:

1. debris damage – mostly from the external tank and the solid rocket boosters at take-off (due to poorly installed/maintained insulation), and space debris;
2. debonding caused by factors other than debris damage – weakness in the tile system (due to inadequate tile placement/maintenance or repeated exposure to load cycles);
3. re-entry heating – re-entry temperature may exceed the capacity of the tiles (i.e. burn-through due to malfunction in guidance system) and may cause damage to the shuttle's systems.

The event-tree shown in Figure A.5 indicates the relationship between tile failure and LOV. Typical event probabilities used in the risk analysis are presented in Table A.6. These event probabilities were obtained from a combination of (1) frequencies of events from official or personal records and (2) subjective assessments. For example, 12 black tiles have

```
┌──────────────┐
│ Debris Damage │
└──────┬───────┘
       ▼
┌──────────────┐    ┌───────────────┐    ┌──────────────┐    ┌──────────────┐
│ Loss of Tile │───▶│ Reentry Heating│───▶│  Subsystem   │───▶│Loss of Shuttle│
└──────▲───────┘    └───────▲────┬──┘    │  Malfunction │    └──────────────┘
       │                    │    │       └──────────────┘
┌──────┴────────────┐ ┌─────┴────▼──┐
│ Debonding Caused by│ │Loss of Additional│
│Factors Other than Debris│ │    Tiles     │
└───────────────────┘ └──────────────┘
```

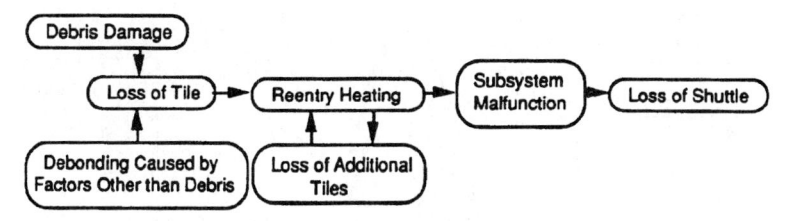

Figure A.5 Event-tree diagram: failure of the thermal protection system leading to loss of vehicle.
Source: Paté-Cornell and Fischbeck (1990).

been found during maintenance to have no bond (apart from the gap filler); this is from a total of 130 000 black tiles installed at various times on all the orbiters. It was then assumed that an equal number of unbonded tiles exists and that these have not yet been detected (i.e. they have shown no visible sign of weakness and hence have not yet been replaced). An expert from the Kennedy Space Centre then estimated that the probability of losing such an unbonded tile was approximately 10^{-2} per flight. Therefore, the probability of losing a single tile due to inadequate tile placement is $12/130\,000 \times 10^{-2} \approx 9.0 \times 10^{-7}$ per flight (see Table A.7). The risk analysis also incorporated the influence of the location of the tiles and their loss on the probability of LOV; this is referred to as 'risk-criticality' (see Figure A.6).

The risk analysis found that the mean probability of LOV (due to failure of the TPS) was 1.2×10^{-3} per flight. NASA estimates an overall probability of LOV (for all causes) of approximately 1×10^{-2} per flight; hence approximately 10% of the overall probability of LOV is due to TPS failure (a not insignificant proportion). Figure A.7 shows that 14% of the tiles accounted for 80% of the risk; it is these tiles that are 'risk-critical'. Further, it was observed that the most 'risk-critical' tiles were not necessarily located in the hottest areas of the orbiter's surface.

It should be clear from the above that a significant proportion of the risk can be reduced by (1) improved inspection of the most 'risk-critical' tiles (i.e. more efficient use of resources) and (2) relating human errors

Table A.7 Typical event probabilities for the thermal protection system of the Space Shuttle Orbiter

Event	Pr(event)/flight
Single tile lost – debris	1.0E-7
– inadequate tile placement/maintenance	9.0E-7
– repeated exposure to load cycles	2.0E-7
Burn-through given that single tile lost	0.1

Source: adapted from Paté-Cornell and Fischbeck (1990).

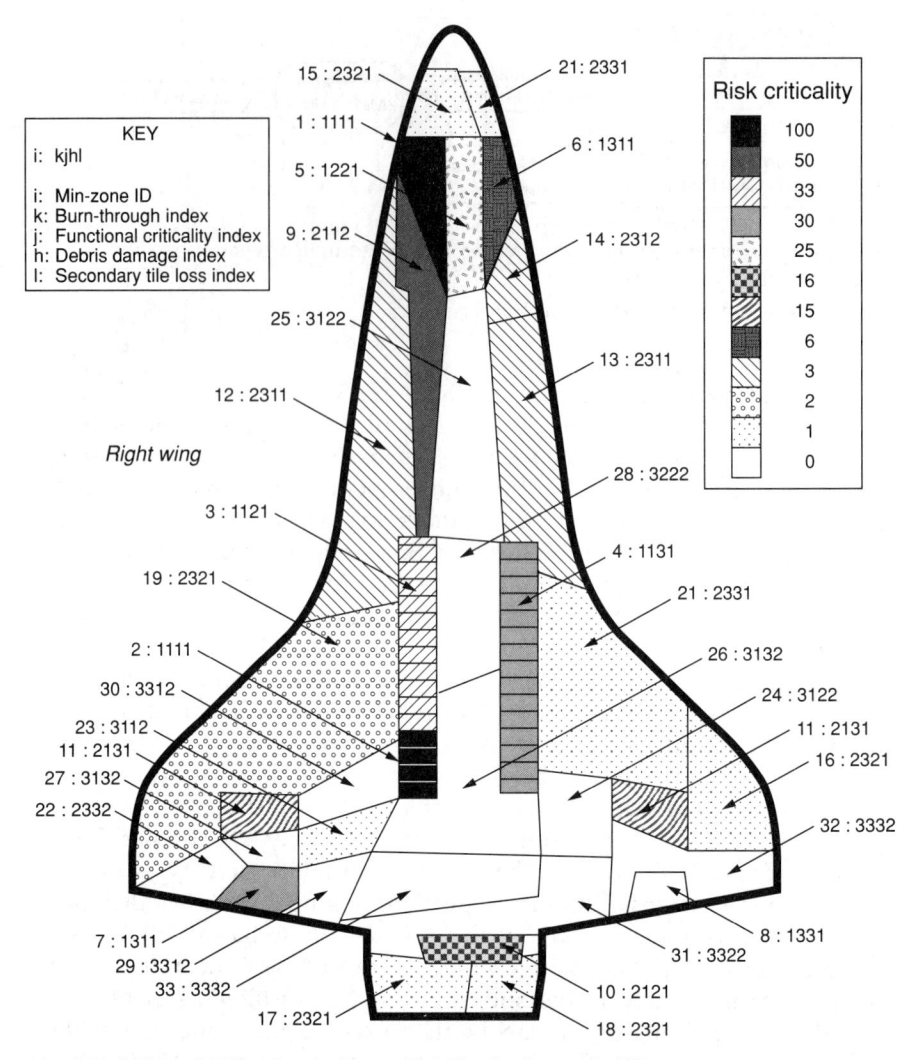

Figure A.6 Partition of the Orbiter's surface into 33 mini-zones.
Source: Paté-Cornell and Fischbeck (1990).

to organizational factors. For these reasons, Paté-Cornell and Fischbeck (1990) identified the following organizational (or management) factors as having an important influence on TPS performance:

1. time pressure (tile maintenance and replacement is often on the critical path of the next launch);
2. liability concerns and conflict among contractors;

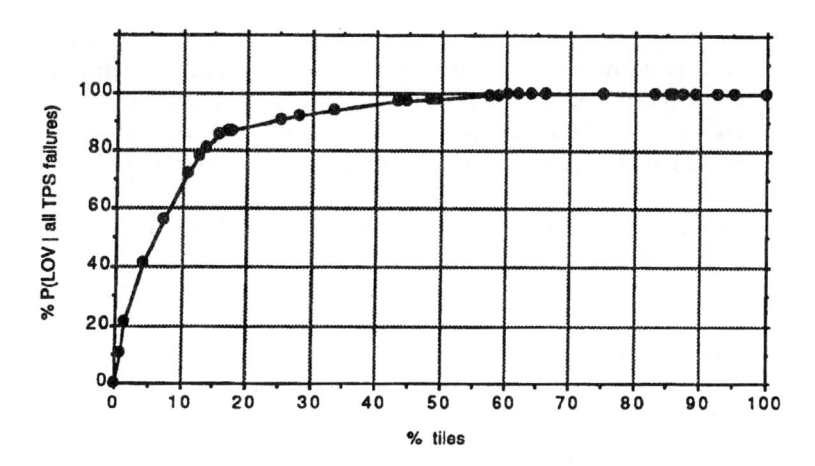

Figure A.7 Risk-criticality of tile loss.
Source: Paté-Cornell and Fischbeck (1990).

3. low status of tile workers – leading to high turnover of experienced staff;
4. random testing;
5. handling of external tank and solid rocket boosters.

These factors provided a basis for further investigation of issues such as identification of management improvements, their cost and their influence on TPS performance.

A.5 'CALIBRATION' – STRUCTURAL RELIABILITY OF TENSION MEMBERS

In Australia, as in a number of other countries, the design codes for structural design have been (or are being) converted from a Working Stress Design (WSD) format to a Limit State Design (LSD or 'load and resistance factor design' – LRFD) format. For steel design the relevant codes are AS1250 (1981) and AS4100 (1990) respectively. The WSD format measures safety in terms of a Factor of Safety (e.g. maximum load divided by minimum resistance – generally a factor of 1.66). However, the LSD format uses probabilistic 'calibration' methods such that designed members have similar probabilities of failure.

A typical design equation for a LSD code provision is

$$\phi R^* \geq \gamma_G G + \gamma_Q Q \tag{A.1}$$

where R^* is the design capacity of the member (e.g. in bending, or tension etc.), G and Q are the design dead and live load effects respectively (obtained from AS1170.1–1989), γ_G and γ_Q are the dead and live load factors, and ϕ is the capacity reduction factor. For various values of γ_G, γ_Q and ϕ estimates can be made of structural reliability for the member. This allows assessment of the consistency of structural reliabilities across a range of member sizes and duties and for a range of dead and live loads.

A convenient measure of structural reliability for 'load-resistance' elements is the safety index (β), calculated as

$$p_f = \Pr(R \leq S) = \Phi(-\beta) \tag{A.2}$$

where p_f is the probability of failure, S is the 'actual' load effect, R is the ultimate structural resistance, and Φ represents the cumulative distribution function of the standard normal distribution. The actual load effect is the sum of $\bar{D}$ and $\bar{L}_p$ where $\bar{D}$ and $\bar{L}_p$ denote the dead load effect and the 50-year maximum live load effect respectively.

Pham (1987) has reported the calculated safety indices (β) for tension members designed in accordance with AS1250 and AS4100 in order to determine the most appropriate values for γ_G, γ_Q and ϕ. Both codes require that two possible modes of failure be considered:

1. yielding of gross section; and
2. net fracture of the section at the connection.

Probabilistic models for actual resistances (R) and loads ($\bar{D}$ and $\bar{L}_p$) are given in Table A.8.

The safety indices obtained from the calibration process for the WSD and LSD codes are presented in Figure A.8, for $\gamma_G = 1.25$, $\gamma_Q = 1.5$ and $\phi = 0.9$. Note that the dead to live load ratios are generally between 0.2 and 0.6 for most steel structures. It is observed from Figure A.8 that the safety indices all exceed a target safety index of 3.0, usually considered appropriate for dead and live loadings (Ellingwood et al., 1980). Figure A.8 shows also that there is considerable scatter of safety indices for the WSD provisions. However, there is less scatter for the LSD provisions,

Table A.8 Statistical parameters for load and resistance parameters

Parameter	G/G+Q	Mean	Coefficient of variation	Type of distribution
LOADS:				
$\bar{D}/G$	–	1.05	0.10	Lognormal
$\bar{L}_p/Q$	0.0–0.4	0.71–0.90	0.25	Gumbel
	0.5–1.0	0.96	0.25	Gumbel
RESISTANCE:				
R/R*: (i) Gross section Failure	–	1.17	0.09	Lognormal
(ii) Net Fracture	–	1.16	0.08	Lognormal

Source: Pham (1985; 1987).

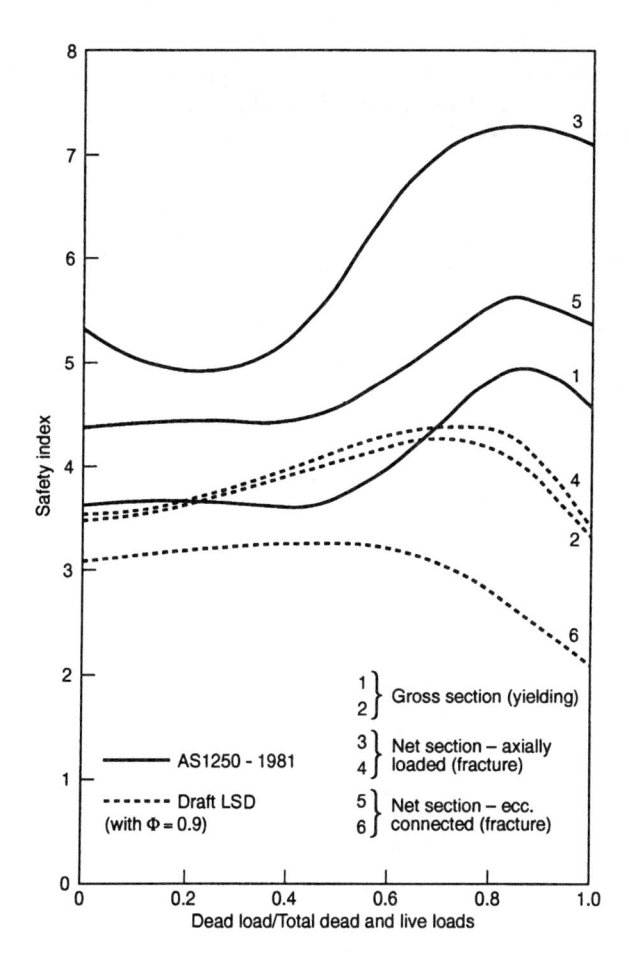

Figure A.8 Safety indices for tension members.
Source: Pham (1987).

which means that the provisions produce more consistent probabilities of failure. This suggests that the design factors used for Figure A.8 are appropriate for all categories of tension members (Pham, 1987).

A.6 GRAVITY DAM

For gravity dams, possible failure scenarios include ageing, persistent overtopping (caused by excessive annual floods), transient overtopping, foundation instability and earthquakes. The most important failure modes usually are persistent overtopping and earthquakes, and in a risk analysis the remaining failure scenarios are often omitted from consideration. System failure may then be defined as the consequences of a sudden release of water. For example, persistent overtopping may scour the dams unprotected foundation, thus causing dam rupture, and hence causing loss of the dam and flood damage to the downstream population.

As usual, event-trees representing failure events and consequences of system failure can be developed for each failure mode. In the risk analysis described by Bury and Kreuzer (1986) failure events were influenced by reservoir level at the time of the flood, the amount of flood warning given to the dam operator, the probability of the operator incorrectly closing the gates (i.e. not initiating a controlled lowering of reservoir), the reliability of the gate mechanism, and antecedent spillway conditions. Failure consequences were affected by such mitigating actions as the amount of warning given to the downstream population and the effectiveness of evacuation procedures. Figures A.9 and A.10 show event-trees describing sequences of failure events and consequences of failure respectively, for persistent overtopping. Similar event-trees were developed for earthquake events. Figures A.9 and A.10 show that the annual probabilities of dam failure ($P_{1.1}$, $P'_{1.1}$ – based on functionality of spillway) and event

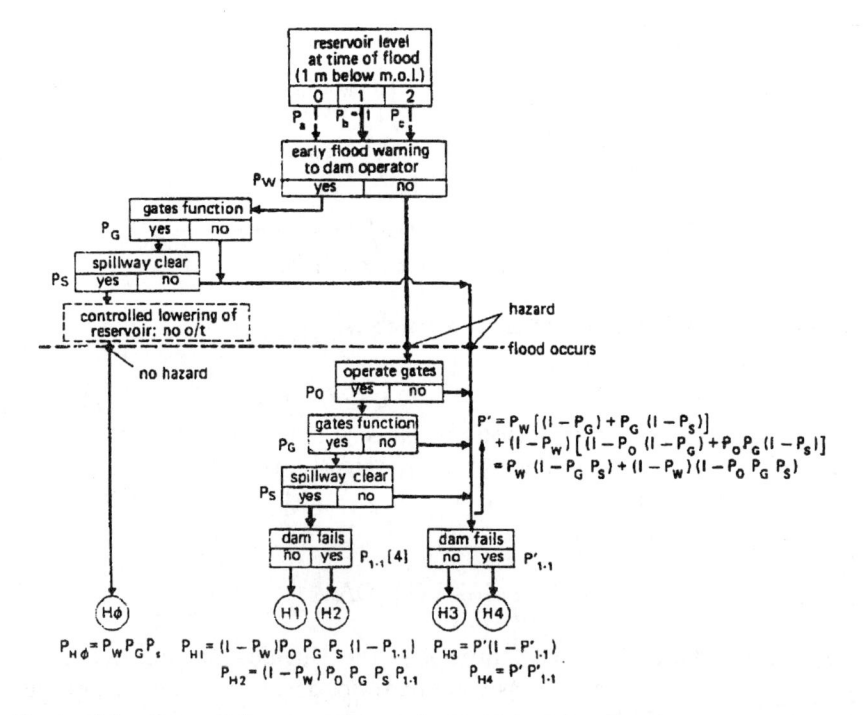

Figure A.9 Cause-failure event tree, for persistent overtopping.
Source: Kreuzer (1986).

Gravity dam

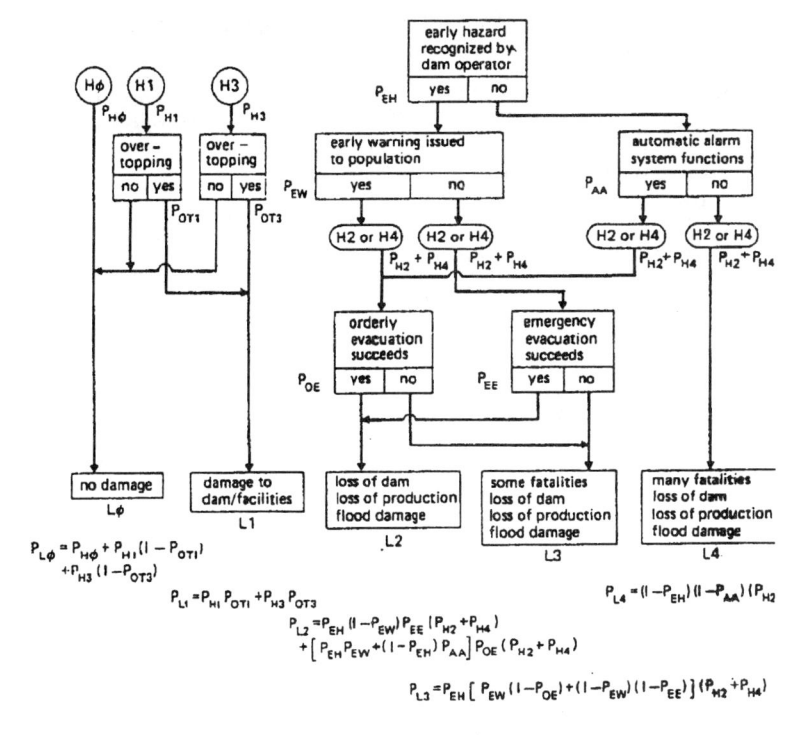

Figure A.10 Failure-consequence event tree, for persistent overtopping.
Source: Kreuzer (1986).

probabilities (e.g. P_O, P_G, P_S, P_{EW}, P_{EE}) are required in order to calculate the annual probabilities of the failure events.

Annual probabilities of dam failure were calculated using 'load-resistance' methods developed to represent:

1. the dam's resistance to sliding (i.e. at its foundation),
2. flood and
3. earthquake loads, see Table A.9.

The models were derived from flood and earthquake frequency event data and from subjective assessments of past experience. Event probabilities were obtained mainly from subjective estimates. Some typical estimates of annual probabilities of dam failure and event probabilities are shown in Table A.10 (Bury and Kreuzer, 1985).

Estimates of monetary losses to be expected for each of the various consequences are given in Table A.11. These loss estimates are subject to some significant uncertainties; including estimating the worth of a human life. The annual probabilities of the corresponding failure events are also given in Table A.11. These probabilities are obtained from the analysis of the failure cause and consequence event-trees. Table A.11

Table A.9 Uncertainties associated with typical input parameters

Parameter	Parameter value	Coefficient of variation
Dam resistance to flooding:		
μR	3406 (t/m)	0.05
σR	329	0.05
Dam resistance to earthquake:		
μR	4017 (t/m)	0.05
σR	394	0.05
Flood load:		
μL	1862 (t/m)	0.08
σL	14.6	0.16
Earthquake load:		
μL	1965 (t/m)	0.05
σL	139	0.45

Source: Bury and Kreuzer (1985,1986).

Table A.10 Typical event probabilities

Event[a]		Probability
$P_{1.1}$	Dam failure	0.9×10^{-6}
$P'_{1.1}$	Dam failure (gates not opened)	7.7×10^{-6}
P_W	Early flood warning given to operator	0.2
$(1-P_G)$	Malfunctioning gates	0.0001
$(1-P_S)$	Clogged spillway	0.0005
$(1-P_O)$	Incorrectly operated gates	0.0005
P_{EW}	Early warning issued to population	0.4
P_{EE}	Emergency evacuation succeeds	0.7

Note: [a] refers to notation used in Figures A.9 and A.10
Source: adapted from Bury and Kreuzer (1986).

also shows the final system risks in terms of dollar amount losses per year; these system risks being computed as the product of the annual probability of occurrence of the failure event and the estimated loss for each consequence event. They may be considered as expected values.

Comparison of the system risks shows that earthquake activity is the dominant source of risk. However, it is unclear if this risk is sufficiently high to warrant the need for risk mitigating actions (e.g. remedial works, improvement in evacuation procedures etc.).

Bury and Kreuzer (1986) also investigated the influence of parameter uncertainty by assuming that the probabilities of dam failure and the event probabilities were all normally distributed random variables. An uncertainty analysis using Monte Carlo simulation found that the upper 75% confidence limit on annual probabilities of dam failure ($P_{1.1}$, $P'_{1.1}$)

Table A.11 Loss and risk estimates

Consequences	Flood				Earthquake			
	Case	Annual Probabilities	Loss $\times 10^6$	Risk[a] $/year	Case	Annual probabilities	Loss $\$\times 10^6$	Risk[a] $/year
Damage to dam/facilities	L1	7.8×10^{-6}	5	40	L6	1.1×10^{-3}	5	5500
Loss of dam/production	–	–	–	–	L7	3.4×10^{-6}	300	1000
… + flood damage	L2	5×10^{-7}	350	180	L8	4.5×10^{-6}	400	1800
… + some fatalities	L3	2×10^{-7}	550	110	–	–	–	–
… + many fatalities	L4	0	750	0	L9	3.4×10^{-6}	800	2700
Total				330				11000

Note: [a] refers to expected value or system risk.
Source: adapted from Bury and Kreuzer (1986).

was an order of magnitude larger than the corresponding mean value. This information and the uncertainties for the event probabilities were included in the analysis used to calculate risk estimates. As a result, measures of uncertainty (i.e. coefficients of variation) in the risk analysis outcome can be obtained. For example, the coefficient of variation for the total earthquake risk ($11 000/yr) was found to be approximately 2.8; this reflects a significant and rather high measure of uncertainty in the outcomes (note that this did not include uncertainties in estimating the monetary losses). It appears that most of the variation was due to uncertainties in the calculation of the annual probabilities of dam failure. Uncertainties in the event probabilities were found to have little influence on this variation.

Finally, Bury and Kreuzer (1986) conclude that

> in spite of the inherent uncertainties, risk assessment is of value for the comparison of several dams, or dam modifications, or dam designs, in terms of their cost/benefit ratio, provided the analysis is consistent for all alternatives. Absolute values of cost/benefit appear less trustworthy, particularly in view of the fact that a residual risk (inadequate models, omissions, gross errors) remains, and is difficult to quantify.

REFERENCES

AS1170.1 (1989) *Minimum Design Loads on Structures: Part 1 – Dead and Live Loads and Load Combinations*, Standards Association of Australia, Sydney.

AS1250 (1981) *Steel Structures Code*, Standards Association of Australia, Sydney.

AS4100 (1990) *Steel Structures Code*, Standards Association of Australia, Sydney.

Boykin, R.F., Freeman, R.A. and Levary, R.R. (1984), Risk assessment in a chemical storage facility, *Management Science*, **30**(4), 512–17.

Bury, K.V. and Kreuzer, H. (1985) Assessing the failure probability of gravity dams, *Water Power and Dam Construction*, **37**(11), 46–50.

Bury, K.V. and Kreuzer, H. (1986) The assessment of risk for a gravity dams, *Water Power and Dam Construction*, **38**(12), 36–40.

Ellingwood, B., Galambos, T.V., MacGregor, J.G. and Cornell, C.A. (1980) *Development of a Probability Based Load Criterion for American National Standard A58*, National Bureau of Standards Special Publication 577, US Government Printing Office, Washington, DC.

Melchers, R.E. (1987) *Structural Reliability: Analysis and Prediction*, Ellis Horwood, Chichester, England.

Okrent, D. (1987) Safety goals, uncertainties, and defense in depth. In Y.Y. Haimes and E.Z. Stakhiv (eds), *Risk Analysis and Management of Natural and Man-Made Hazards*, 268–82.

Paté-Cornell, M.E. and Fischbeck, P.S. (1990) *Safety of the Thermal Protection System of the Space Shuttle Orbiter: Quantitative Analysis and Organizational Factors – Phase 1: Risk-Based Priority Scale and Preliminary Observations*, Report to the National Aeronautics and Space Administration.

Pham, L. (1987) Safety index analyses of tension members, *Civil Engineering Transactions*, **CE29**(2), 128–30.

SYSTEMS RELIABILITY SERVICE, Data Products Group, United Kingdom Atomic Energy Authority, Warrington, England.
USNRC (1989) *Severe Accident Risks: An Assessment for Five Nuclear Power Plants*, NUREG-1150, US Nuclear Regulatory Commission, Washington, DC.

Index

Page numbers appearing in **bold** refer to figures and page numbers appearing in *italics* refer to tables.

Accident sequence trees, *see* Event trees
Accidents, *see* Sources of risk
Acceptance criteria
   outcome from probabilistic risk analysis 168–170
   quantitative risk acceptance criteria 237, **239**
   see also Decision analysis, Regulatory safety goals, Risk perception
Aerospace systems
   human error 30–31
   quantitative safety targets 236
   risk analysis of space shuttle orbiter 256–259
   space vehicles 31
   system failure 29–30
Aircraft, *see* Aerospace systems
ALARA 228, 233
   see also Regulatory safety goals
ALARP 228, **228**
   see also Regulatory safety goals
Applications
   calibration of structural design 259–261
   chemical storage facility 252–256
   dams 261–266
   nuclear power plants 249–252
   space shuttle orbiter 256–259

Bathtub curve 81–82, **82**, 180
Bayes theorem 88–90, **90**

Cause tree, *see* Fault tree
Chemical plants
   accident databases 51
   component reliability databases 91
   HAZOP 44, 50–51
   human error 22
   performance shaping factors 129–130
   quantitative safety targets 234–236
   risk analysis of chemical storage facility 252–256
   system failure 21
Common cause failures 65–69, 174–179
   dependency matrix 174, **175**
   dependency modelling 66, 68, 174–179
   evaluation 174–179
   event and fault trees 68, 174, **177**
   example 66
Common mode failures, *see* Common cause failures
Computer systems 32
Concrete strength 102
Consequences 115–116
Consequence trees, *see* Event trees
Contingency measures 133–134
Corrosion 102
Cost benefit analysis, *see* Decision analysis

Dams
   human error 28
   loads 28
   resistance 28
   risk analysis of dam 261–266
   system failure 28
Decision analysis
   attributes 216–217, 218, *225*
   cost-effectiveness 221
   economic value of human life 219–220
   expected utility analysis 216, 221–227

expected value analysis 216, 218–221, 227
limitations 221, 226
multi-attribute utility 224–225
objectives 216–217
risk-cost-benefit analysis 218, 219
supra decision maker 216
time 221, 226
utility functions 222–223, **222**, **223**, **224**
Decision tree 54, 63
see also Event tree
Dependency modelling, *see* Common cause failures
Deterministic variable 69, 157
limitations 70
see also Quantified risk analysis (QRA)

Earthquakes, *see* Loads
Error factor 137–139, **139**
Errors, *see* Human error
Expected utility analysis, *see* Decision analysis
Expected value analysis, *see* Decision analysis
Event tree 54–56, 60–65
combination with fault trees 54, 61, **63**, 162, **163**
common cause failures 68
cut set 57–59, 61–63
decision tree 54, 63
evaluation 159–161, 165–166, 167, 171–173, 183–185
examples **63**, **64**, **160**, **168**
human reliability analysis 123
minimal cut set 57–59, 61–63
Monte-Carlo simulation 171–173
reduced or truncated 61, **65**, 173–174
Expert opinion 86–88, 88–90, 107, 136, 157–158

Failure rate, *see* Reliability data
Fault tree 54–60
basic events 56
boolean logic 56–57
combination with event trees 54, **63**, 162, **163**
common cause failures 68, **177**
cut set 57–59
evaluation 161–162, 166–167, 183–185
examples **58**, **59**, **162**

limitations 60
minimal cut set 57–59
for offshore platform 54–55
parallel system 57, 183–185, **184**
reduced or truncated 173–174, **250**
series system 57, 183–185, **184**
FMEA, *see* Failure modes and effect analysis
FMECA, *see* Failure mode, effect and criticality analysis
Failure
definition of failure 42
environmental damage 43
human error 122
recovery from 35
Failure modes and effect analysis (FMEA) 44, 46–49, 50
Failure mode, effect and criticality analysis (FMECA) 44, 49–50
First order reliability (FOR) method 196–197
First order second moment (FOSM) method 190–196
bounds 194–195
example 194
limit state function 190, 193
limitations 196, 197
Floods, *see* Loads
Fragility curve 95–97, **96**

HAZOP, *see* Hazard and operability studies
Hazard and operability studies (HAZOP) 44, 50–51
Hazard function, *see* Reliability data
High consequence – low probability events 13, 115, 220
Human error 3, 36, 122–149
cause of failure 122, *123*
classification 124–129
concrete workmanship 102
definition 124
error control 129–130, 132–134
human reliabilities 69, 134–149
organisational errors 127–128, *128*
performance shaping factors 129–130, *129*, 133, 135
trivial errors 132
unfore*see*n errors 131–132
violations 130–131, *131*
see also Human reliability data
Human reliability data 134–149
databases 136–149
error magnitude 134

Index

error rate 134
error recovery 134
existing literature 146, *147*
expert opinion 136
HEART 140–142
INTENT 143
MAPPS 145
other databases 145–146
PROF 142–143
SLIM 144
sources 134–136
TESEO 140, *141*
THERP 137–140, *138*
TRC 144
uncertainties 135, 147–148
validation of databases 147–148
Human reliability analysis (HRA) 123, 155, 249
see also Human error, Human reliability data, THERP
Hydrological systems 29

Industrial robots, *see* Robots

Likelihood function 89
Limit state function 98–100, 190–196
see also First order second moment (FOSM) method, Load-resistance sub-systems
Live load 112–113
Load-resistance sub-systems
evaluation 185–200
examples 98, 185–186
first order reliability (FOR) method 196–197
first order second moment (FOSM) method 190–196
loads 105–115
nominal capacity 104
probability of failure 98–100, 185–200
random variables 99
resistance 100–105
see also Resistance, Loads
Loads 25, 28, 69
combinations 114–115
dead 114
earthquakes 106–109
expert opinion 107, *109*
extraordinary 112–113
floods 108–110
live 112–113
other 114

sustained 112
Turkstra's rule 114
wind 110–112

Management 11
Mean value method 104
Metal fatigue 102
Modelling error 104, **105**, 109–110,
Monte-Carlo simulation analysis
advantages 170
computational times 170
dependency 170–171
evaluation of event tree 171–173
importance sampling 198–199
resistance 104
simple (crude) methods 197–198
system evaluation 170–173, **171, 172**, 197–199
variance reduction 173, 198–199

Nuclear power plants
accident databases 51
complex interactions 34–35
component reliability databases 91
earthquake *21*
performance shaping factors 129–130
probabilistic risk analysis 249–253
quantitative safety targets 233–234
system failure 17
valve failure 18–19, *20*

Offshore platforms
component reliability databases 91
fault tree 54–55
human error 28
incident databases 51
quantitative safety targets 236
system failure 27

Peer review 241–243
Perception of risk, *see* Risk perception
Performance shaping factors 129–130, 133, 135
Pipelines 24–25
Point estimate, *see* Deterministic variable
Posterior distribution 89
Preliminary hazard analysis 44–46
Prior distribution 88–89
Probability 3–4, 155–157
conditional 156

distribution 72–74, **74**
uncertainty 157
Probability of failure 98–100, 185–200
  see also System evaluation
Probabilistic risk analysis (PRA) 4, 115, 155, 164–174
  common cause failures 174–179
  event trees 165–166, 167, **168**
  first order reliability (FOR) method 196–197
  first order second moment (FOSM) method 190–196
  fault trees 166–167
  nuclear power plant 249–252
  objective risk 209
  outcome of analysis 168–170, **169**, 239–240
  random variables 164–165
  reliability of load-resistance sub-systems 185–200
  second moment analysis 165–170
  sensitivity analysis 200
  simplification of system 173–174
  time effects 179–185
  uncertainty 165
  updating 200
  see also Monte-Carlo simulation analysis
Probabilistic safety analysis (PSA), see Probabilistic risk analysis (PRA)
Probability distribution 72–74, **74**
  see also Random variable
Probable maximum flood 110
Probable maximum precipitation 110

Quality assurance 12, 74, 241–243, 242, **242**
Qualitative risk analysis 157–158, *159*
Quantified risk analysis (QRA) 155, 158–163
  combination 162, **163**
  common cause failures 174–179
  event trees 159–161, **160**
  fault trees 161–162, **162**
  limitations 157, 169, 239
  objective risk 209
  reliability of load-resistance sub-systems 185–200
  sensitivity analysis 200
  simplification of system 173–174
  time effects 179–185
  updating 200

  see also Monte-Carlo simulation analysis
Quantitative safety targets, see Regulatory safety goals

Random variable 69, 70–74
  coefficient of variation 73
  continuous 72–73
  discrete 72–73
  mean 73
  probabilistic model 72–74, 86
  resistance 100–105
  standard deviation 73
  stochastic processes 106
  uncertainty 70, 73–74, 157
  variability 70, 105
  variance 73
  see also Load-resistance sub-systems, Loads, Probabilistic risk analysis (PRA), Resistance
Regulatory safety goals
  ALARA 228, 233
  ALARP 228, **228**
  calibration 232, 259–261
  deemed to comply regulations 229–231
  individual fatality risk 236
  issues 237–241
  litigation 227
  quantifiable performance requirements 229, 231–232
  quantitative safety targets 229, 232–241, *235*, **238**, **253**
  safety case 229–230
  societal risk **238**
  specification standards 229, 230–231
  types of regulations 227–229
  uncertainties 239–241
Reliability block diagram **59**
Reliability data
  combining different data 88–90
  component failure data 79, 83–97
  databases 90–95, *91*, *92*, *93*, *94*
  expert opinion 86–88
  external factors 95–97
  field data 84
  failure rate 79–82, 83, 180–181
  hazard function 180, **181**
  hazard rate 182
  incident databanks 84
  laboratory testing 83–84
  load-resistance sub-systems 80
  mean time to failure 79, 181–183

Index

repair times 80, 83
system evaluation 183–185
time effects 179–185
unavailability 80, 82–83
uncertainties 84–86
variation 81–82
see also Human reliability data
Resistance 25, 28, 69, 100–105
Risk
  acceptable 208
  definition 3
  individual vs. societal 4–5, **210**
  tolerable 208
  legal implications 13–14
  voluntary vs. involuntary 4–5
  see also Sources of risk
Risk acceptance criteria, see Acceptance criteria
Risk analysis
  component reliability 79–97
  uncertainty 74
  see also Applications, Human reliability analysis, Probabilistic risk analysis, Qualitative risk analysis, Quantified risk analysis
Risk assessment 1–2, 9–10
  context definition 5–7
  decision-making 2–3, 206–207
  decision analysis 216–227
  hazard identification 7–8
  perception of risk 208–216
  politics 12
  public participation 12–13
  quality assurance 241–243
  regulatory safety goals 227–241
  risk analysis 8–9
  risk management 2, **6**, 204
  risk treatment 10–11
  sensitivity analysis 9, 200
  see also Acceptance criteria, Decision analysis, Regulatory safety goals, Risk perception
Risk communication 214–215
Risk management 2, **6**, 204
  see also Acceptance criteria, Risk analysis, Risk assessment
Risk perception
  acceptable risk 208
  comparative risks 210–211, *211*
  culture 214
  discussion 215–216
  government spending 212–213, *213*
  objective risk 209
  psychological aspects 209–213
  risk communication 214–215
  socio-demographics 213–214
  societal risk **210**
  tolerable risk 208–209
Robots 31

Safety case 229–230
  see also Regulatory safety goals
Safety goals, see Regulatory safety goals
Serviceability failure 25
Seismic hazard curve 107, **108**
Sensitivity analysis 9, 200
Shipping
  accident database 51
  human error 23
  quantitative safety targets 236
  system failure 23
Simulation, see Monte-Carlo simulation analysis
Sources of risk
  aerospace systems 29–31
  chemical plants 21–23
  computer systems 32
  dams 28–29
  failure modes and effect analysis (FMEA) 44, 46–49, 50
  failure mode, effect and criticality analysis (FMECA) 44, 49–50
  general causes 17
  hazard and operability studies (HAZOP) 44, 50–51
  hydrological systems 29
  identification of risks 43–54
  incident databanks 51–54
  nuclear power plants 17–21
  offshore platforms 27–28
  pipelines 24–25
  preliminary hazard analysis 44–46
  robots 32
  recovery from failure 35
  shipping 23–24
  structures 25–27
  unfore*seen* events 32–34
Space shuttle orbiter 256–259
Space vehicles, see Aerospace systems
Statistical parameters 72–73
Steel strength 100–101
Stochastic process 106
Structures
  accident database 54

human error 26
loads 25
resistance 25
system failure 25
System elements, *see* Load-resistance sub-systems, Reliability data
System evaluation 154–155
see also Quantified risk analysis (QRA), Probabilistic risk analysis (PRA)
System failure, *see* Failure, System evaluation
System representation, *see* Fault trees, Event trees
System risk, *see* Risk, System evaluation

Task Driven Method 47
THERP 137–140, *138*

Time effects 179–185
Toxicology 116
Turkstra's rule 114

Unexpected events, *see* Unforeseen events
Unforeseen events 32–34
complex interactions 34–35
errors 131–132
Utility theory 221–223

Variables
uncertainty 70
see also Deterministic variable, Random variable
Violations 130–131
see also Human error

Wind, *see* Loads